SCOTTISH CERTIFICATE OF EDUCA

Higher
PHYSICS

The Scottish Certificate of Education Examination Papers
are reprinted by special permission of
THE SCOTTISH QUALIFICATIONS AUTHORITY

Note: The answers to the questions do not emanate from the Authority

ISBN 0 7169 9288 4

ROBERT GIBSON · Publisher
17 Fitzroy Place, Glasgow, G3 7SF.

CONTENTS

DATA SHEET

COMMON PHYSICAL QUANTITIES

Quantity	*Symbol*	*Value*	*Quantity*	*Symbol*	*Value*
Speed of light in vacuum	c	$3{\cdot}00 \times 10^{8}$ m s^{-1}	Mass of electron	m_e	$9{\cdot}11 \times 10^{-31}$ kg
Charge on electron	e	$-1{\cdot}60 \times 10^{-19}$ C	Mass of neutron	m_n	$1{\cdot}675 \times 10^{-27}$ kg
Gravitational acceleration	g	$9{\cdot}8$ m s^{-2}	Mass of proton	m_p	$1{\cdot}673 \times 10^{-27}$ kg
Planck's constant	h	$6{\cdot}63 \times 10^{-34}$ J s			

REFRACTIVE INDICES

The refractive indices refer to sodium light of wavelength 589 nm and to substances at a temperature of 273 K.

Substance	*Refractive index*	*Substance*	*Refractive index*
Diamond	2·42	Glycerol	1·47
Glass	1·51	Water	1·33
Ice	1·31	Air	1·00
Perspex	1·49		

SPECTRAL LINES

Element	*Wavelength*/nm	*Colour*
Hydrogen	656	Red
	486	Blue-green
	434	Blue-violet
	410	Violet
	397	Ultraviolet
	389	Ultraviolet
Sodium	589	Yellow

Element	*Wavelength*/nm	*Colour*
Cadmium	644	Red
	509	Green
	480	Blue

Lasers

Element	*Wavelength*/nm	*Colour*
Carbon dioxide	9550, 10590	Infrared
Helium-neon	633	Red

PROPERTIES OF SELECTED MATERIALS

Substance	*Density*/ kg m^{-3}	*Melting Point*/ K	*Boiling Point*/ K	*Specific Heat Capacity*/ J kg^{-1} K^{-1}	*Specific Latent Heat of Fusion*/ J kg^{-1}	*Specific Latent Heat of Vaporisation*/ J kg^{-1}
Aluminium	$2{\cdot}70 \times 10^{3}$	933	2623	$9{\cdot}02 \times 10^{2}$	$3{\cdot}95 \times 10^{5}$	....
Copper	$8{\cdot}96 \times 10^{3}$	1357	2853	$3{\cdot}86 \times 10^{2}$	$2{\cdot}05 \times 10^{5}$	....
Glass	$2{\cdot}60 \times 10^{3}$	1400	....	$6{\cdot}70 \times 10^{2}$	....	....
Ice	$9{\cdot}20 \times 10^{2}$	273	....	$2{\cdot}10 \times 10^{3}$	$3{\cdot}34 \times 10^{5}$	....
Glycerol	$1{\cdot}26 \times 10^{3}$	291	563	$2{\cdot}43 \times 10^{3}$	$1{\cdot}81 \times 10^{5}$	$8{\cdot}30 \times 10^{5}$
Methanol	$7{\cdot}91 \times 10^{2}$	175	338	$2{\cdot}52 \times 10^{3}$	$9{\cdot}9 \times 10^{4}$	$1{\cdot}12 \times 10^{6}$
Sea Water	$1{\cdot}02 \times 10^{3}$	264	377	$3{\cdot}93 \times 10^{3}$	....	....
Water	$1{\cdot}00 \times 10^{3}$	273	373	$4{\cdot}19 \times 10^{3}$	$3{\cdot}34 \times 10^{5}$	$2{\cdot}26 \times 10^{6}$
Air	1·29	....	....	....	....	....
Hydrogen	$9{\cdot}0 \times 10^{-2}$	14	20	$1{\cdot}43 \times 10^{4}$	....	$4{\cdot}50 \times 10^{5}$
Nitrogen	1·25	63	77	$1{\cdot}04 \times 10^{3}$	....	$2{\cdot}00 \times 10^{5}$
Oxygen	1·43	55	90	$9{\cdot}18 \times 10^{2}$	....	$2{\cdot}40 \times 10^{5}$

The gas densities refer to a temperature of 273 K and a pressure of $1{\cdot}01 \times 10^{5}$ Pa.

SCOTTISH
CERTIFICATE OF
EDUCATION

PHYSICS
HIGHER GRADE
Paper I

SECTION A

For questions 1 to 30 in this section of the paper, an answer is recorded on the answer sheet by indicating the choice A, B, C, D or E by a stroke made in ink in the appropriate box of the answer sheet—see the example below.

EXAMPLE

The energy unit measured by the electricity meter in your home is the

A ampere

B kilowatt-hour

C watt

D coulomb

E volt.

The correct answer to the question is B—kilowatt-hour. Record your answer by drawing a heavy vertical line joining the two dots in the appropriate box on your answer sheet in the column of boxes headed B. The entry on your answer sheet would now look like this:

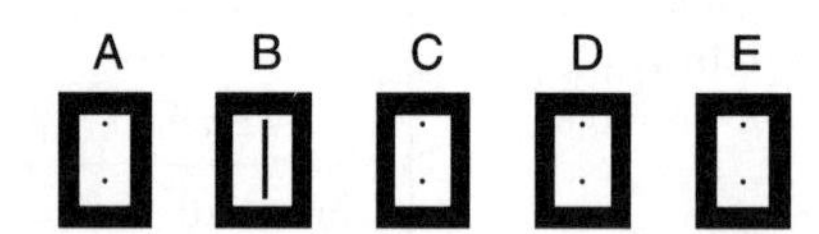

If after you have recorded your answer you decide that you have made an error and wish to make a change, you should cancel the original answer and put a vertical stroke in the box you now consider to be correct. Thus, if you want to change an answer D to an answer B, your answer sheet would look like this:

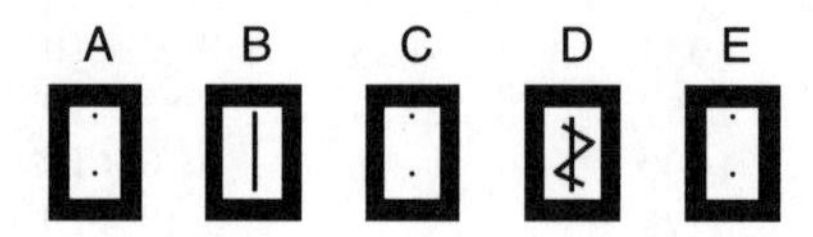

If you want to change back to an answer which has already been scored out, you should enter a tick (✓) to the RIGHT of the box of your choice, thus:

A B C D E OR A B C D E

SCOTTISH CERTIFICATE OF EDUCATION 1994

FRIDAY, 13 MAY
9.30 AM – 11.00 AM

PHYSICS
HIGHER GRADE
Paper I

Read Carefully

1 All questions should be attempted.

2 The following data should be used when required unless otherwise stated.

Speed of light in vacuum c	$3{\cdot}00 \times 10^{8}$ m s^{-1}	Planck's constant h	$6{\cdot}63 \times 10^{-34}$ J s
Charge on electron e	$-1{\cdot}60 \times 10^{-19}$ C	Mass of electron m_e	$9{\cdot}11 \times 10^{-31}$ kg
Acceleration due to gravity g	$9{\cdot}8$ m s^{-2}	Mass of proton m_p	$1{\cdot}67 \times 10^{-27}$ kg

Section A (questions 1 to 30)

3 Check that the answer sheet is for Physics Higher I (Section A).

4 Answer the questions numbered 1 to 30 on the answer sheet provided.

5 Fill in the details required on the answer sheet.

6 Rough working, if required, should be done only on this question paper, or on the first two pages of the answer book provided—**not** on the answer sheet.

7 For each of the questions 1 to 30 there is only **one** correct answer and each is worth 1 mark.

8 Instructions as to how to record your answers to questions 1–30 are previously given.

Section B (questions 31 to 37)

9 Answer questions numbered 31 to 37 in the answer book provided.

10 Fill in the details on the front of the answer book.

11 Enter the question number clearly in the margin of the answer book beside each of your answers to questions 31 to 37.

12 Care should be taken **not** to give an unreasonable number of significant figures in the final answers to calculations.

SECTION A

Answer questions 1–30 on the answer sheet.

1. Consider the following three statements made by pupils about scalars and vectors.

I Scalars have direction only.

II Vectors have both size and direction.

III Speed is a scalar and velocity is a vector.

Which statement(s) is/are true?

A I only

B I and II only

C I and III only

D II and III only

E I, II and III

2. A golfer strikes a ball straight down the fairway.

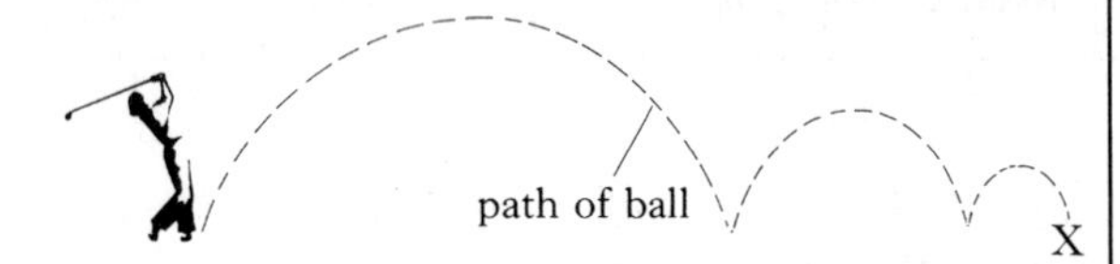

The ball bounces twice before stopping at point X.

Which of the following could be a graph of the **vertical** component of its velocity against time **after** it is struck?

A
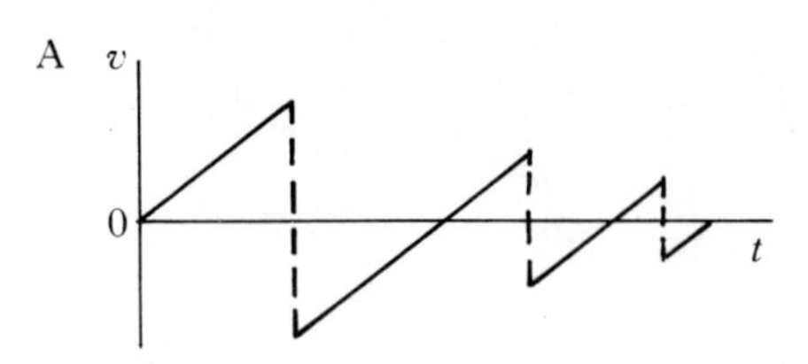

B
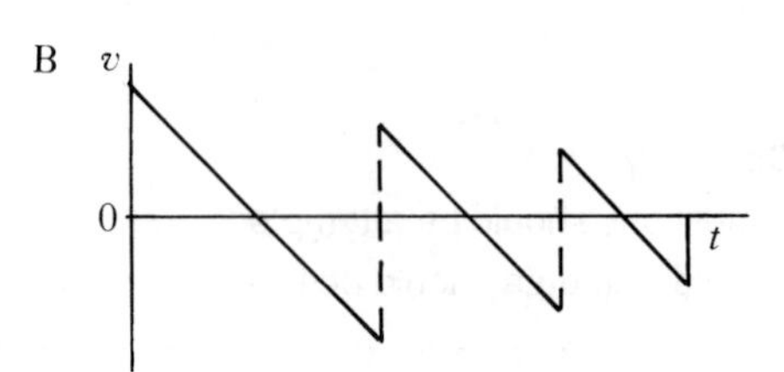

C
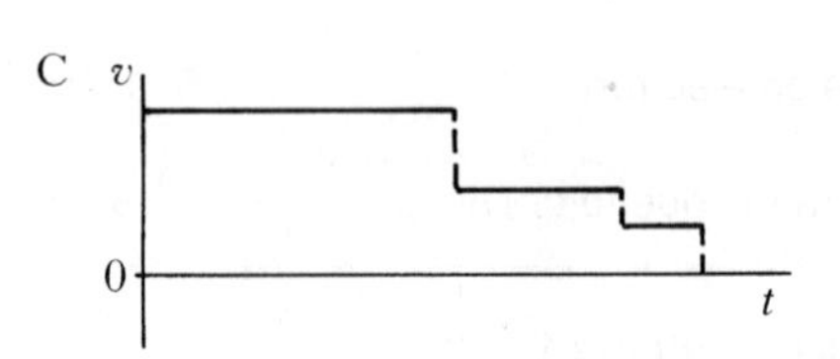

D
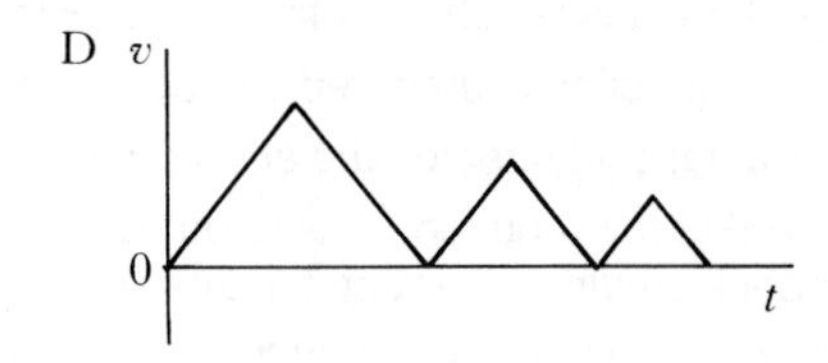

E
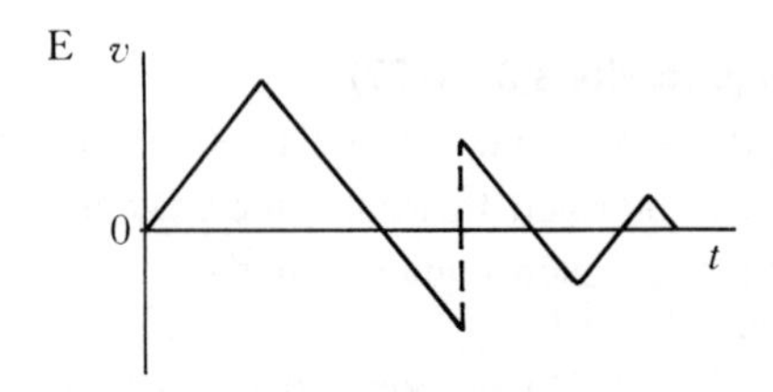

3. A car travelling at 30 m s^{-1} starts to brake when it is 50 m from a stationary lorry. The car moves in a straight line and manages to stop just before reaching the lorry.

What is the deceleration of the car, in m s^{-2}?

A 0·6

B 4·5

C 9

D 10

E 18

4. A lift is raised and lowered by means of a cable.

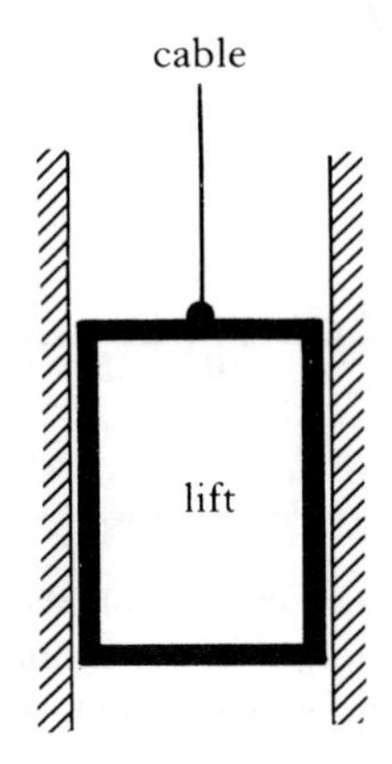

In which of the following situations is the tension in the cable greatest?

A The lift is travelling upwards at a constant speed.

B The lift is travelling downwards at a constant speed.

C The lift is decelerating on the way down.

D The lift is accelerating on the way down.

E The lift is decelerating on the way up.

5. A car of mass 900 kg pulls a caravan of mass 400 kg along a straight, horizontal road with an acceleration of 2 m s^{-2}.

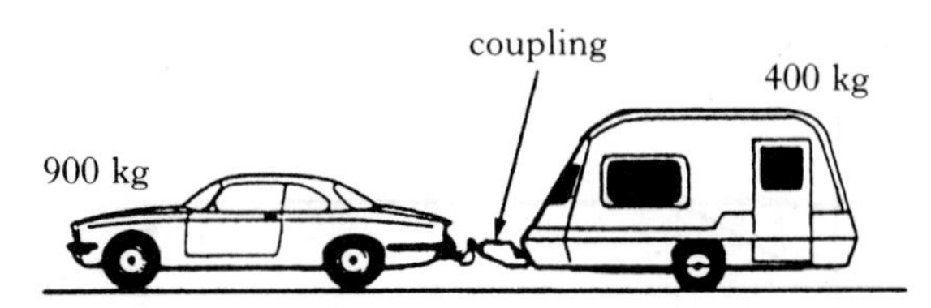

Assuming that the frictional forces are negligible, the tension in the coupling between the car and the caravan is

A 400 N

B 500 N

C 800 N

D 1800 N

E 2600 N.

6. A cyclist free-wheels down a slope, inclined at 15° to the horizontal, at a constant velocity of 3 m s^{-1}.

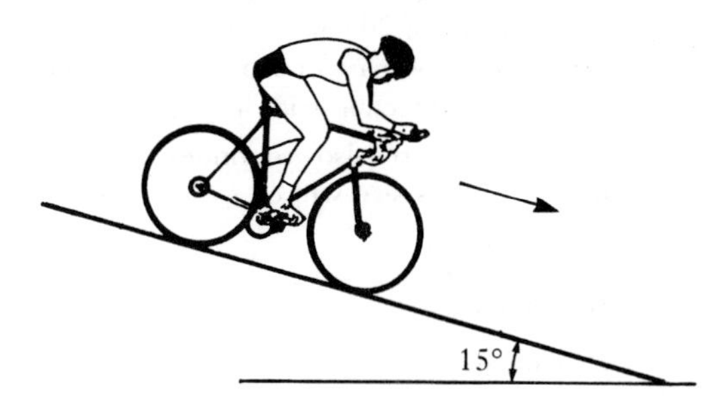

The combined mass of the rider and bicycle is 70 kg. If the value of the acceleration due to gravity is taken as 10 m s^{-2}, the total force of friction is

A 181 N

B 210 N

C 362 N

D 391 N

E 676 N.

7. A force, which is applied in a straight line to an object, varies with time as shown in the following graph.

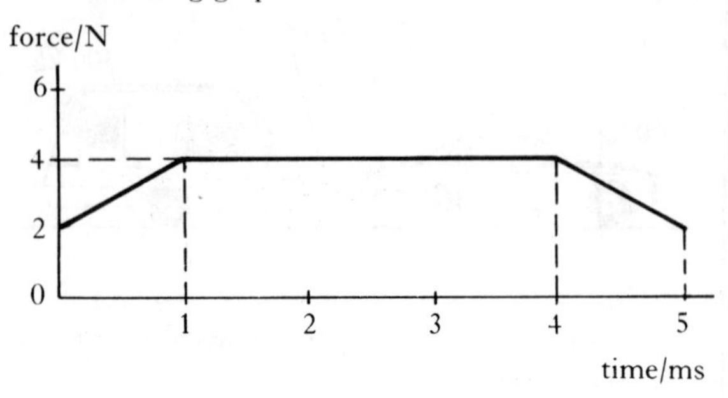

What is the total impulse given to the object by the force in this 5 millisecond time interval?

A 8×10^{-3} N s

B 10×10^{-3} N s

C 15×10^{-3} N s

D 18×10^{-3} N s

E 20×10^{-3} N s

8. A shell of mass 5 kg is travelling horizontally with a speed of 200 m s^{-1} when it explodes into two parts. One part of mass 3 kg continues in the original direction with a speed of 100 m s^{-1}.

The other part also continues in this same direction. Its speed will be

A 150 m s^{-1}

B 200 m s^{-1}

C 300 m s^{-1}

D 350 m s^{-1}

E 700 m s^{-1}.

9. An object of mass 4 kg falls from a considerable height in an area where the acceleration due to gravity is 10 m s^{-2}.

The velocity-time graph for the first 35 seconds of its motion is as follows.

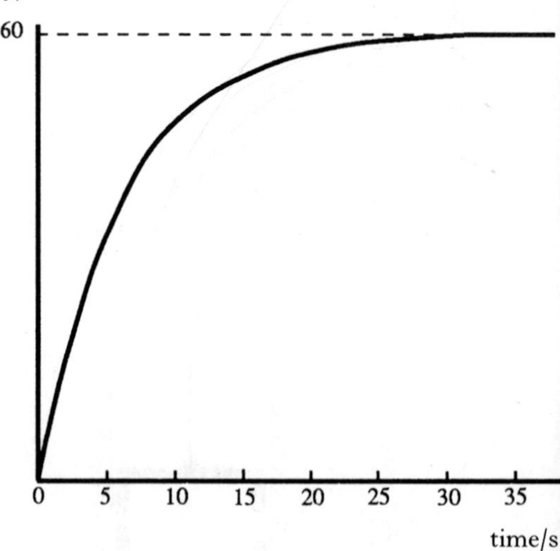

Which row in the following table could give the frictional forces acting on the object at 4 seconds, 8 seconds and 32 seconds?

	Force at 4 s	*Force at 8 s*	*Force at 32 s*
A	0 N	30 N	40 N
B	40 N	30 N	0 N
C	40 N	40 N	40 N
D	20 N	30 N	40 N
E	0 N	0 N	40 N

10. After a car has been parked in the sun for some time, it is found that the pressure in the tyres has increased. This is because

A the volume occupied by the air molecules in the tyres has increased

B the force produced by the air molecules in the tyres acts over a smaller area

C the average spacing between the air molecules in the tyres has increased

D the increased temperature has made the air molecules in the tyres expand

E the air molecules in the tyres are moving with greater kinetic energy.

11. In the arrangement shown below, 2 C of positive charge is moved from plate S, which is at a potential of 250 V, to plate T, which is at a potential of 750 V.

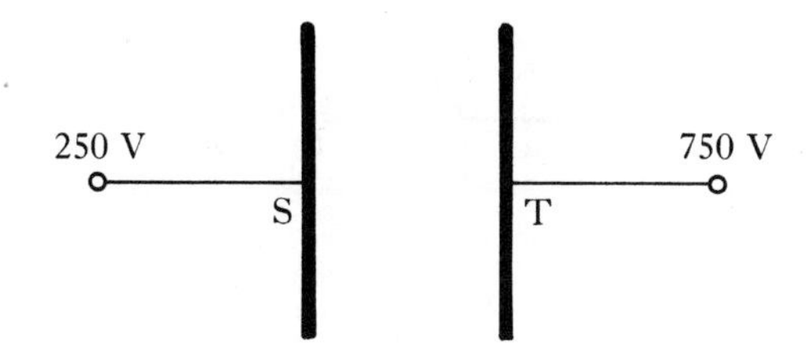

How much energy is required to move this charge from plate S to plate T?

A 0·004 J

B 250 J

C 500 J

D 1000 J

E 1500 J

12. In which of the following arrangements of resistors is the resistance between X and Y the same?

1.
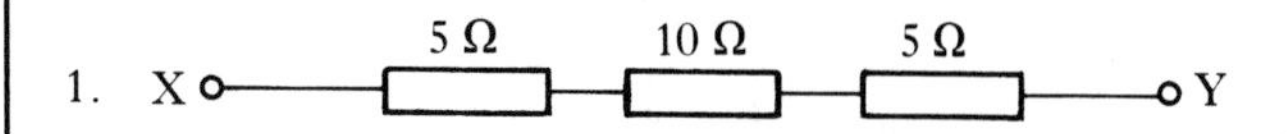

2.
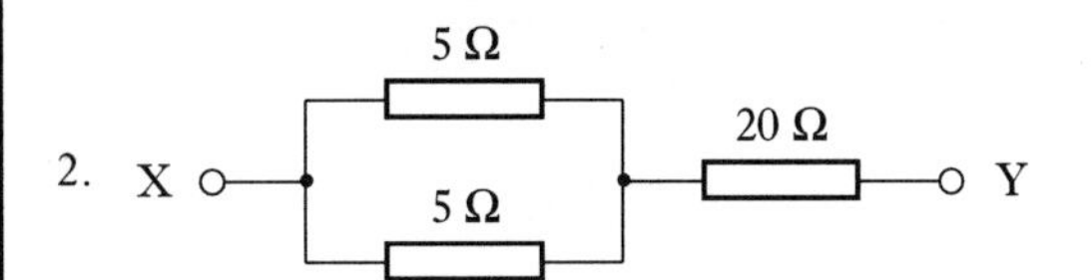

3.
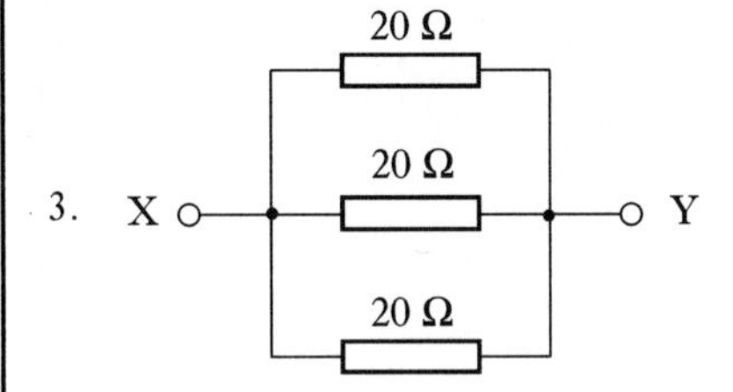

4.
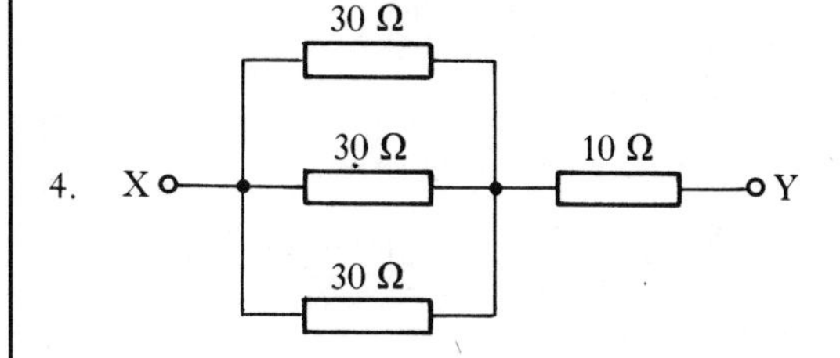

5.
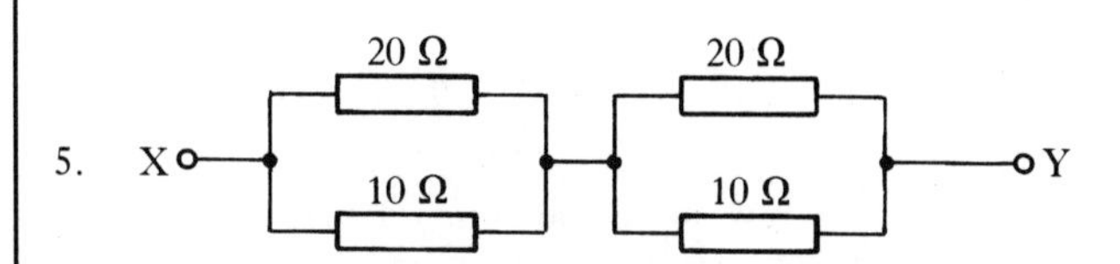

A 1 and 2 only

B 1 and 3 only

C 1 and 4 only

D 1, 2 and 4 only

E 1, 3 and 5 only

13. The current delivered by the battery in the following circuit is 3 A.

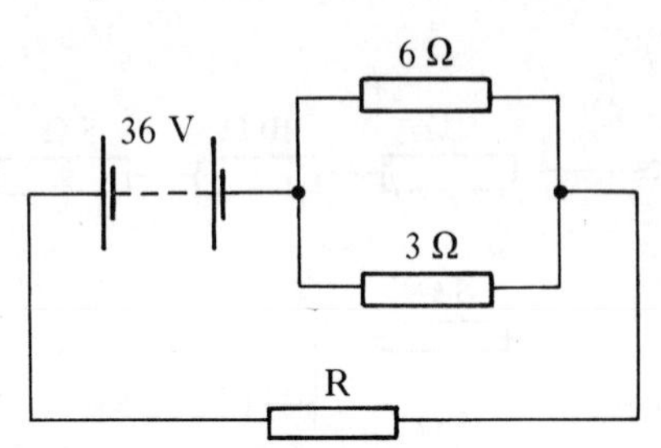

Assuming that the battery has negligible internal resistance, the resistance of resistor R is

A 3 Ω

B 4 Ω

C 10 Ω

D 12 Ω

E 18 Ω.

14. A battery of e.m.f. 12 V and internal resistance 1 Ω is connected across a 2 Ω resistor, as shown in the circuit below.

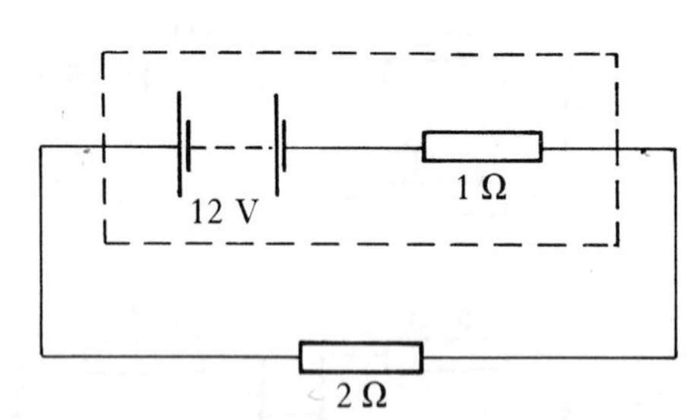

Which row in the following table shows the correct values for current, terminal potential difference and lost volts in this circuit?

	Current/A	*t.p.d./V*	*lost volts/V*
A	4	4	8
B	4	8	4
C	6	4	8
D	6	8	4
E	12	8	4

15. In the following Wheatstone bridge circuit, the reading on the voltmeter is zero when the resistance R_v of the variable resistor is set at 1 kΩ.

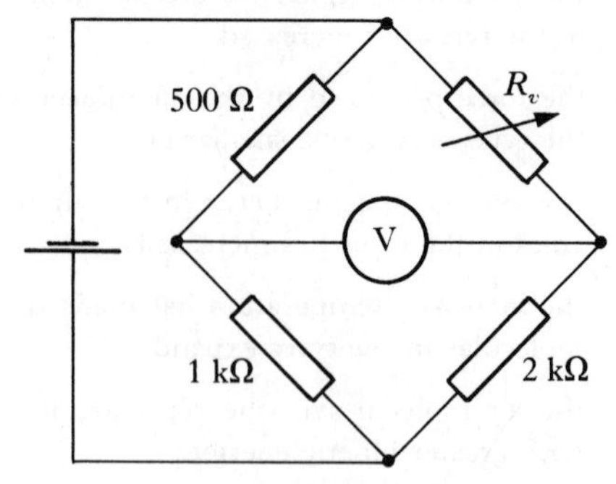

Which of the following would best represent the shape of a graph of the voltmeter reading V against the resistance R_v as it is varied between 990 Ω and 1010 Ω?

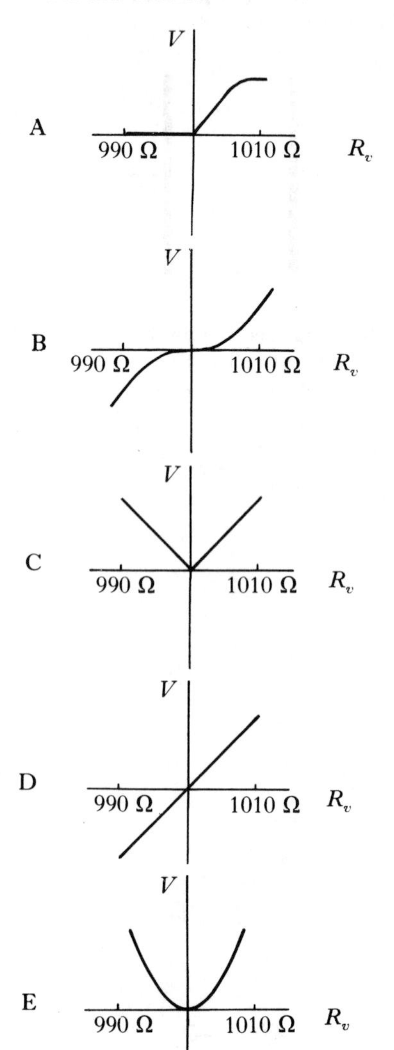

16. The circuit in the diagram below is used for charging a capacitor.

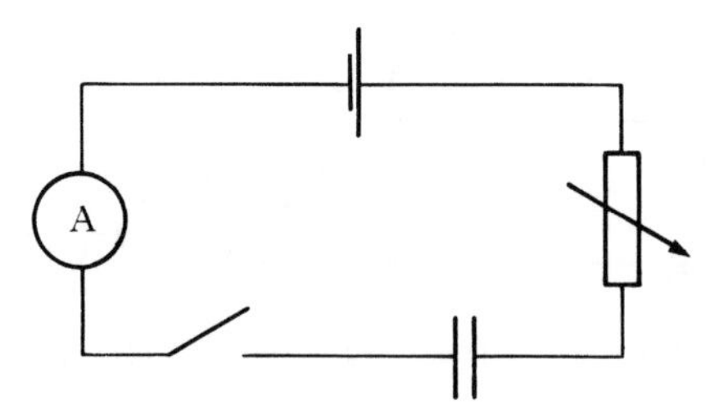

The switch is closed and the capacitor charges up. The variation of current I with time t for this circuit is shown in the following graph.

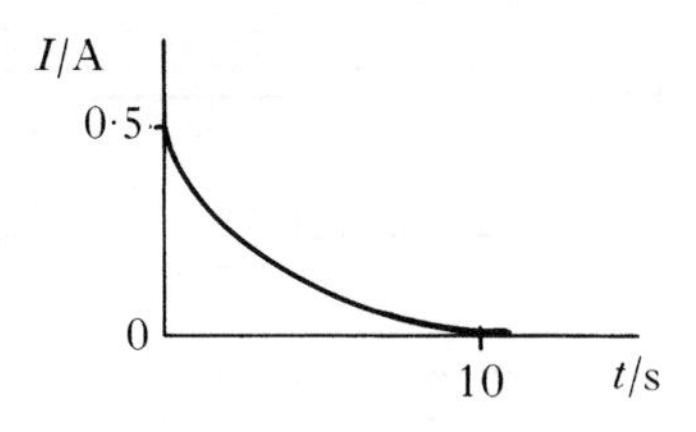

The capacitor is discharged and the value of the variable resistor is **increased.** The experiment is then repeated. Which of the following graphs shows the correct variation of current with time?

A

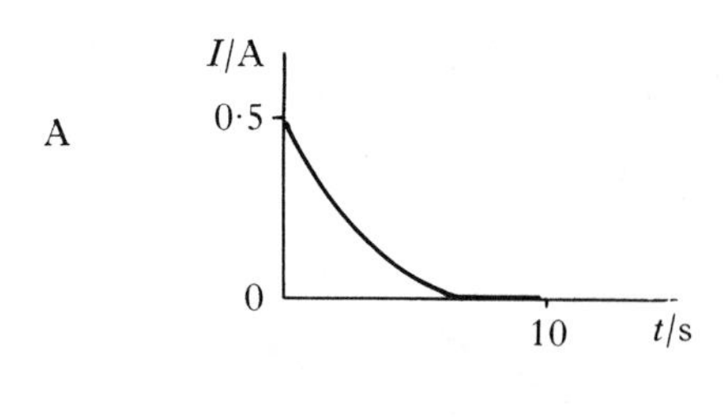

B

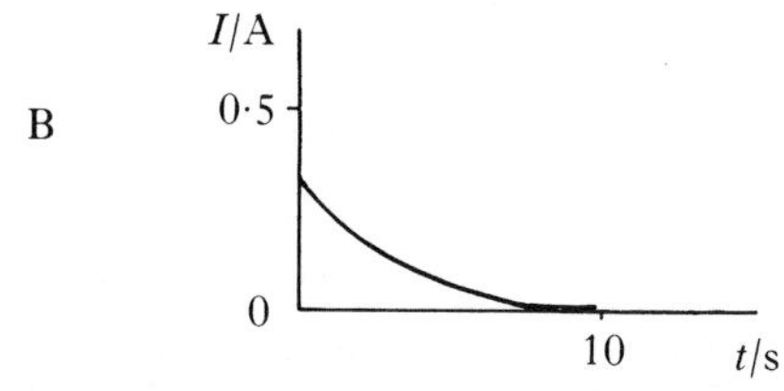

C

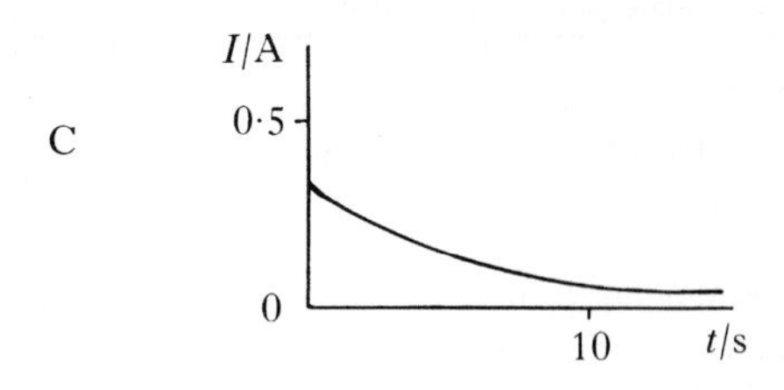

D

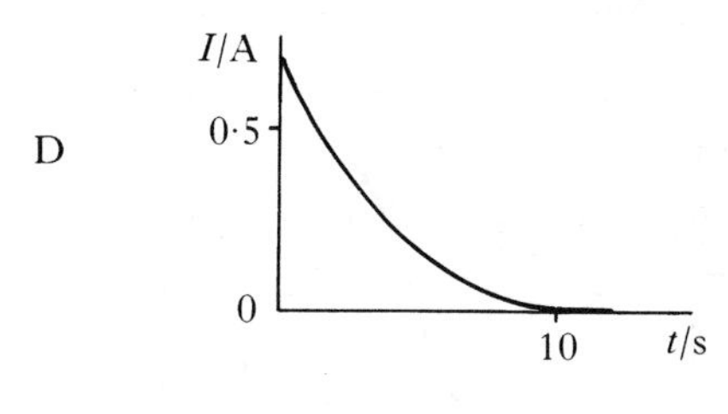

E

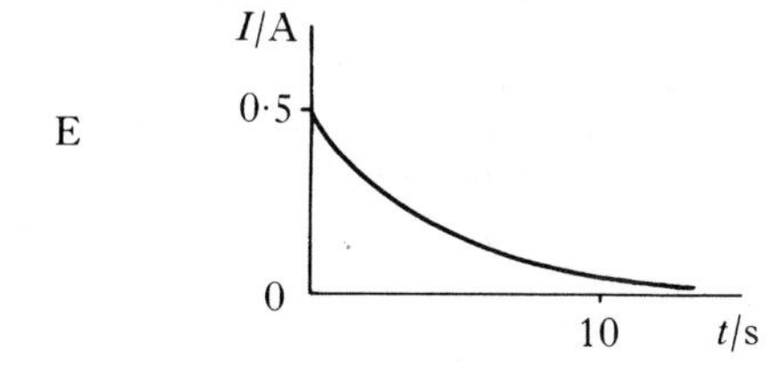

17. In the graph below, the shaded area is used to calculate the work done in charging a capacitor.

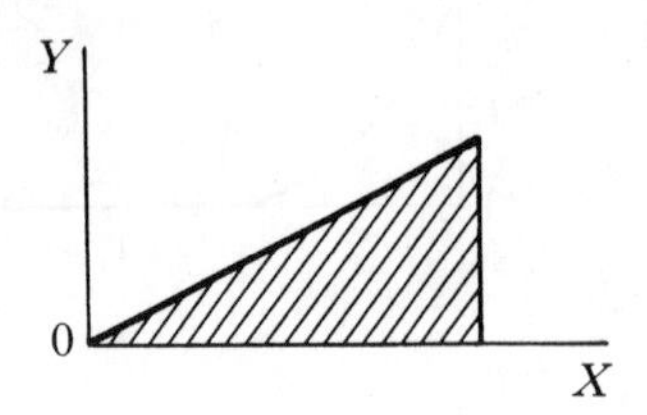

What should be the labels on the X and Y axes?

	X-axis label	*Y-axis label*
A	charge	potential difference
B	current	potential difference
C	charge	time
D	current	time
E	current	charge

18. An immersion heater can be operated either from an a.c. supply or a d.c. supply. The graph below represents the a.c. supply voltage.

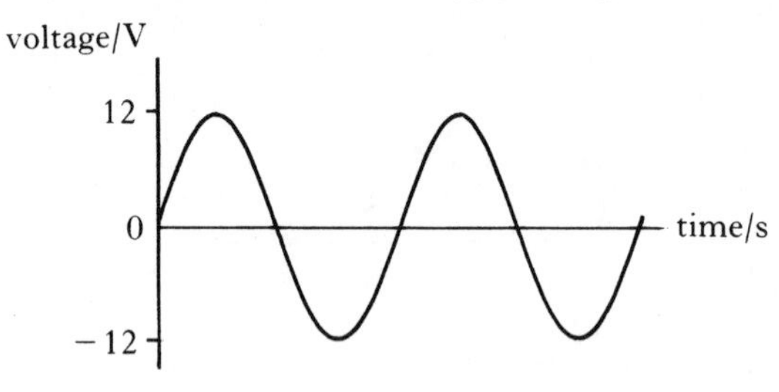

What d.c. supply voltage would produce the same rate of heating from this heater?

A 6 V

B $\frac{12}{\sqrt{2}}$ V

C 12 V

D $12\sqrt{2}$ V

E 24 V

19. Two logic gates are connected together in the following way.

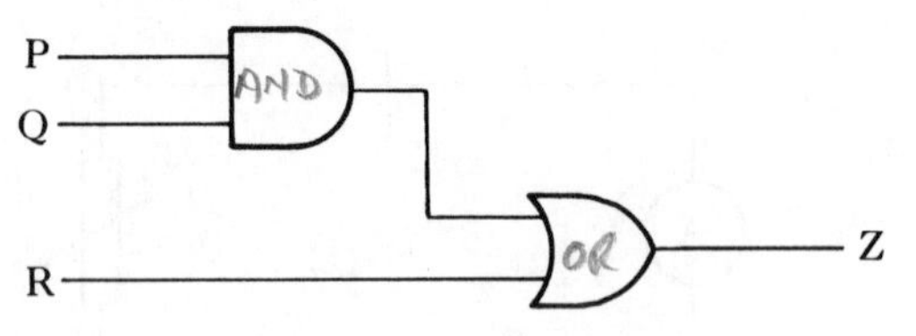

Pulse waveforms are applied to inputs P, Q and R as follows.

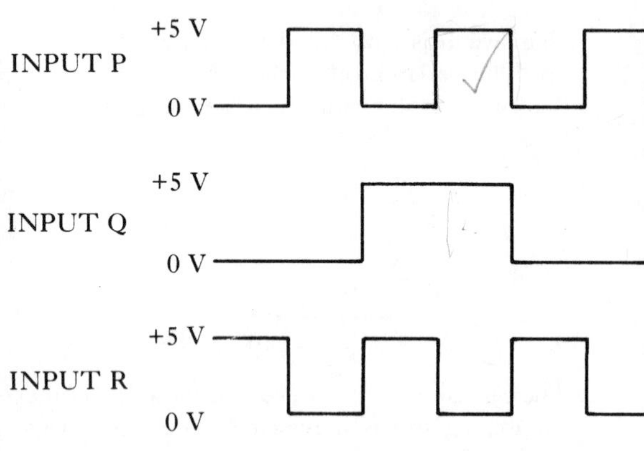

Which of the following shows the output pulse waveform at Z?

A

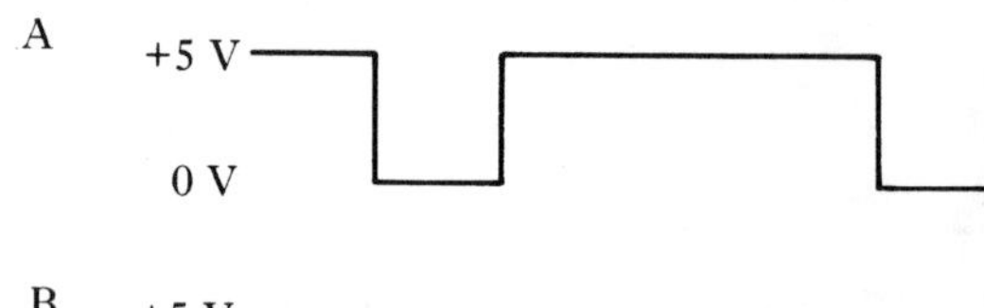

B

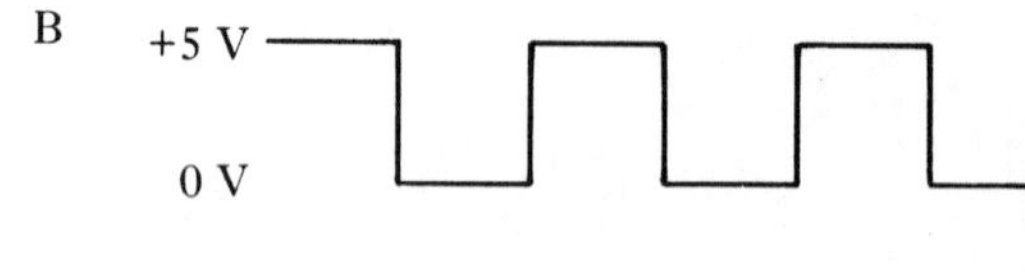

C

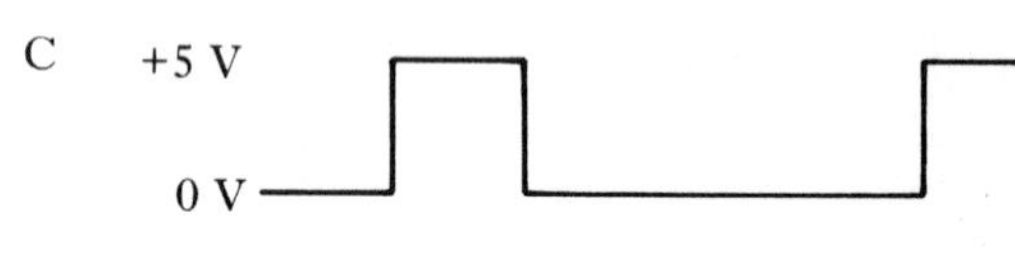

D

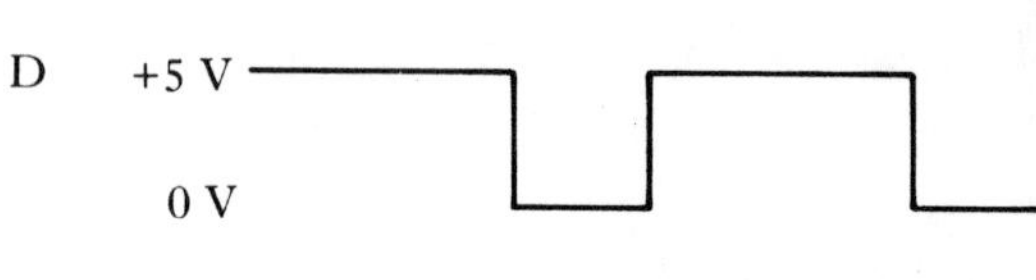

E

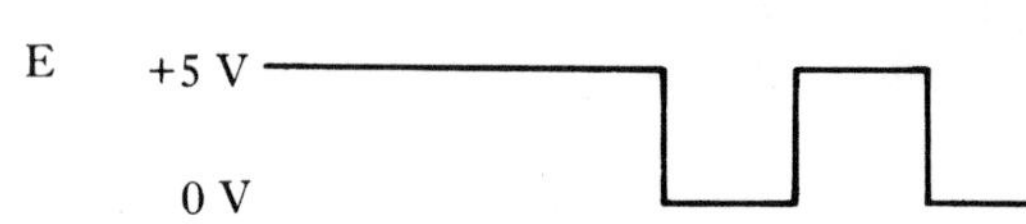

20. The circuit below is used to generate square waves.

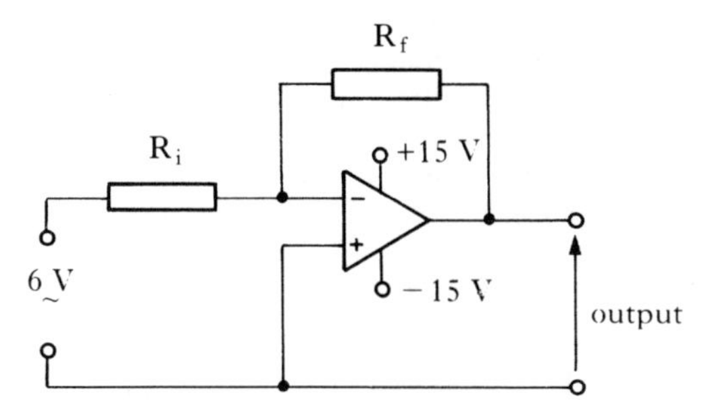

Which values for resistors R_i and R_f will produce an approximately square wave output?

	R_i	R_f
A	1 k Ω	10 k Ω
B	5 k Ω	10 k Ω
C	10 k Ω	10 k Ω
D	10 k Ω	5 k Ω
E	10 k Ω	1 k Ω

21. A ray of light travels with speed v_1 through medium 1 and then passes into another medium 2, where it travels at speed v_2.

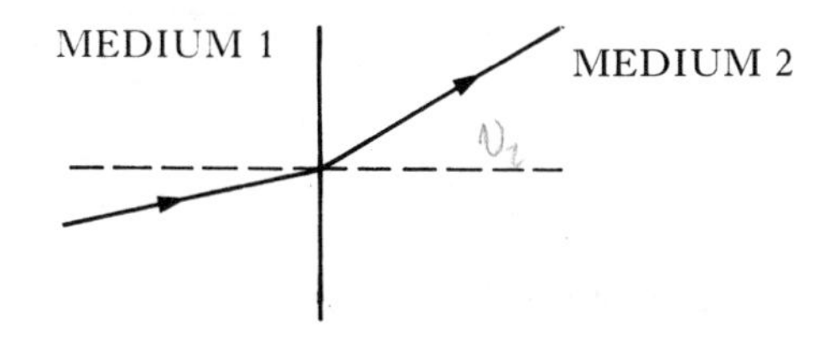

The refractive indices for medium 1 and medium 2 are n_1 and n_2 respectively.

Which row in the following table correctly compares the speeds and refractive indices for each medium?

	Speed of light	*Refractive Index*
A	v_2 is less than v_1	n_2 is less than n_1
B	v_2 is the same as v_1	n_2 is less than n_1
C	v_2 is the same as v_1	n_2 is greater than n_1
D	v_2 is greater than v_1	n_2 is less than n_1
E	v_2 is greater than v_1	n_2 is greater than n_1

22. A pupil sets up the apparatus shown below to investigate the relationship between the angle of incidence (i) and the angle of refraction (r) for a ray of light passing from air into glass.

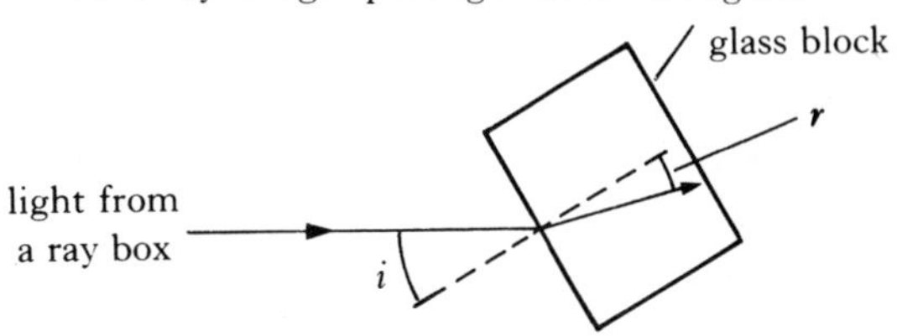

The pupil plots a graph of sin i against sin r.

Which graph shows the correct relationship between sin i and sin r?

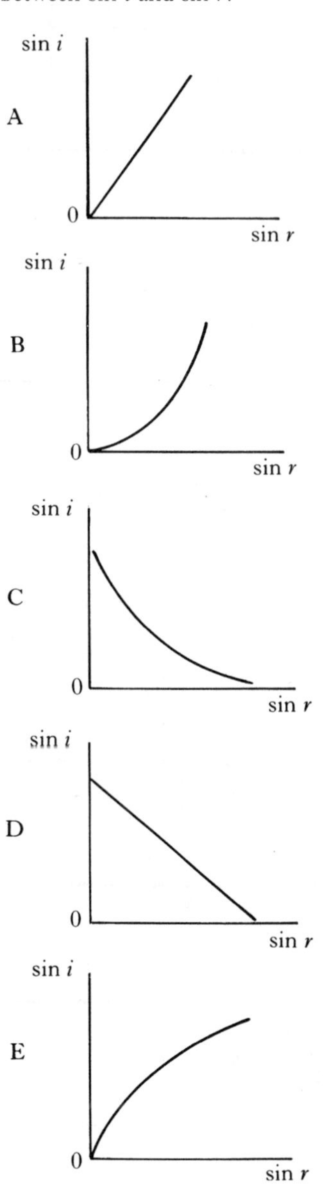

23. A ray of monochromatic light, travelling in air, strikes the side of a rectangular block of glass of refractive index 1·7, as shown below.

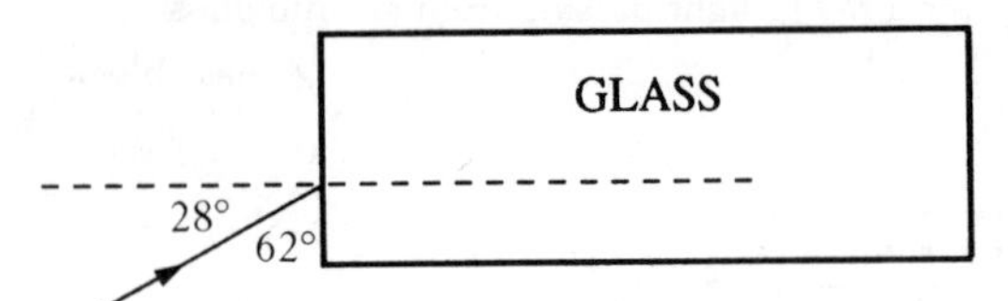

Which of the following diagrams shows correctly the subsequent path of the ray?

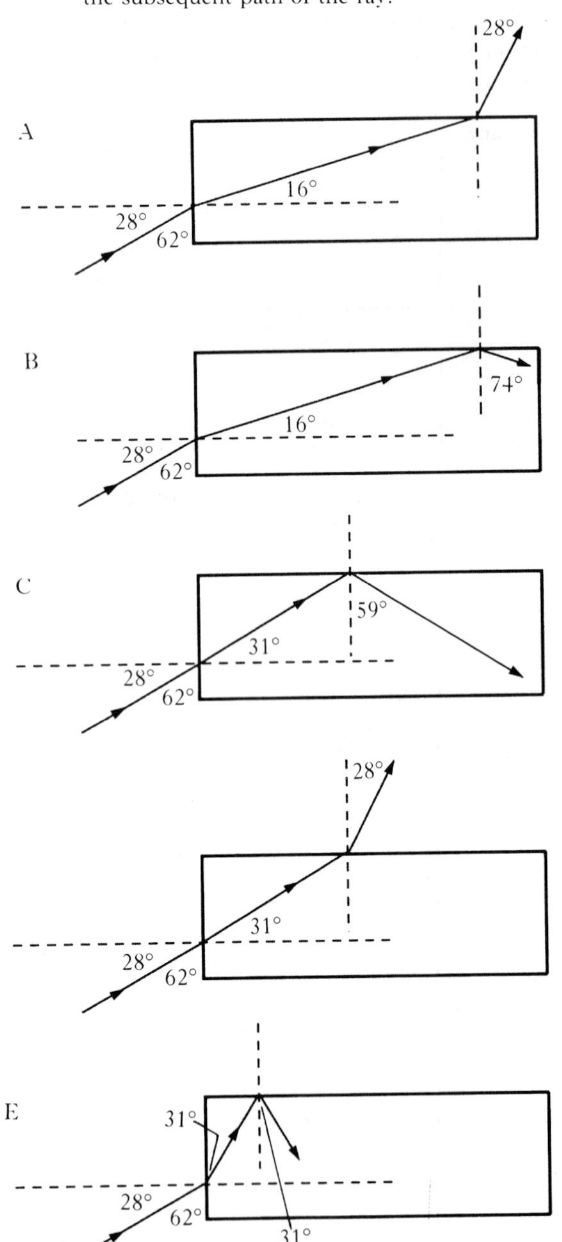

24. Monochromatic light of wavelength λ passes through a grating and produces a pattern of bright maxima on a screen. The separation of lines on the grating is d and the grating is at a distance L from the screen. Which of the following pairs of changes will **always** produce an **increase** in the spacing of the maxima on the screen?

A	increase L	increase d
B	increase λ	increase d
C	decrease L	decrease λ
D	increase L	decrease λ
E	increase λ	decrease d

25. An experiment is carried out to investigate the relationship between the light intensity I from a point source and the distance d from the source. The experiment is done in a darkened room and a meter connected to a light sensor indicates the intensity, as shown below.

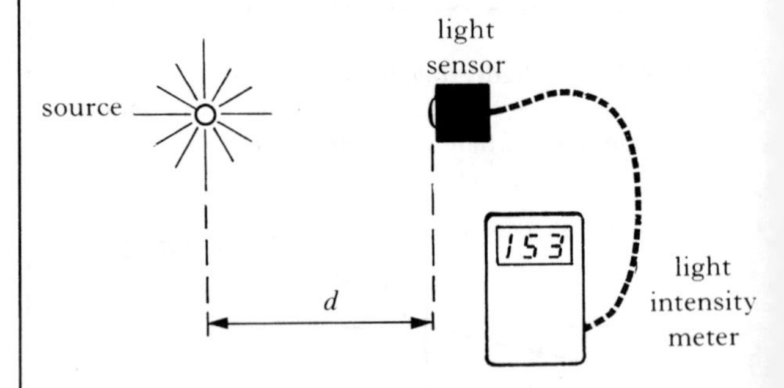

Which of the following expressions will give an approximately constant value?

A $I \times d$

B $I \times d^2$

C $\frac{I}{d}$

D $\frac{I}{d^2}$

E $I \times \sqrt{d}$

26. A student makes a note of the following statements after a lesson about photoelectric emission.

I Photoelectric emission from a metal occurs only if the frequency of the incident radiation is greater than the threshold frequency.

II The threshold frequency depends on the metal from which photoemission takes place.

III If the frequency of the incident radiation is less than the threshold frequency, increasing its intensity will cause photoemission.

Which of the above statements is/are correct?

A I only

B II only

C I and II only

D II and III only

E I, II and III

27. The photon energies for three different radiations are as follows.

Radiation 1: $2{\cdot}78 \times 10^{-19}$ J
Radiation 2: $4{\cdot}97 \times 10^{-19}$ J
Radiation 3: $6{\cdot}35 \times 10^{-19}$ J

Which one of the following is true?

A The wavelength of radiation 1 is longer than that of radiation 2.

B The wavelength of radiation 3 is longer than that of radiation 2.

C The frequency of radiation 1 is higher than that of radiation 2.

D The frequency of radiation 1 is higher than that of radiation 3.

E The frequency of radiation 2 is higher than that of radiation 3.

28. A student reads the following passage in a physics dictionary.

"... a solid state device in which positive and negative charge carriers are produced by the action of light on a p–n junction."

The passage describes a

A light emitting diode

B laser

C capacitor

D photodiode

E thermistor.

29. For the nuclear disintegration described below, which row of the table shows the correct values of x, y and z?

$$^{214}_{x}\mathrm{Pb} \rightarrow {}^{y}_{83}\mathrm{Bi} + {}^{0}_{z}e$$

	x	y	z
A	84	214	1
B	83	210	4
C	85	214	2
D	82	214	−1
E	82	210	−1

30. Three materials X, Y and Z are used as gamma ray absorbers. They have half-value thicknesses of 2 cm, 4 cm and 8 cm respectively.

Gamma rays of intensity I strike the left side of this "sandwich" composed of X, Y and Z.

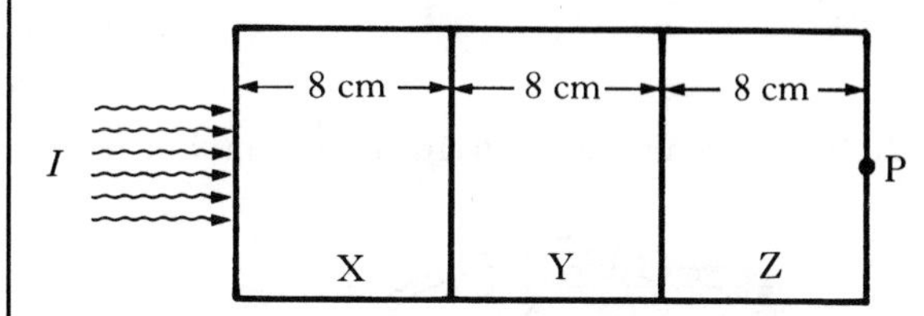

The intensity at point P, on the right side of the "sandwich", will be

A $\frac{I}{8}$

B $\frac{I}{16}$

C $\frac{I}{32}$

D $\frac{I}{64}$

E $\frac{I}{128}$.

SECTION B

Write your answers to Questions 31–37 in the answer book.

Marks

31. The velocity–time graph shown below is for an object moving with constant acceleration a.

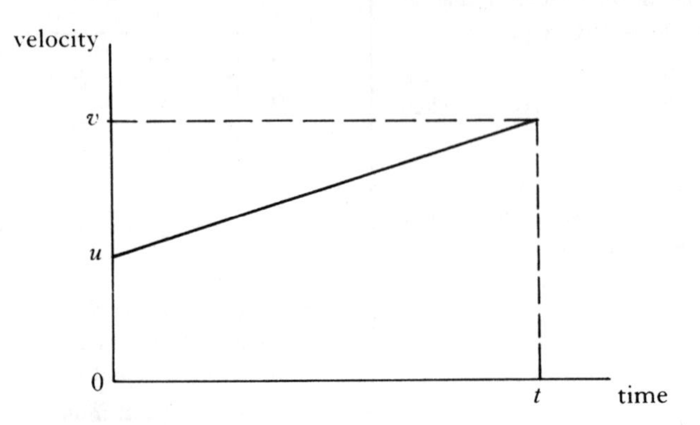

Show that during the time interval t the object moves through a displacement s given by

$$s = ut + \frac{1}{2}at^2.$$ **2**

32. The apparatus in the diagram below may be used to measure the density of air.

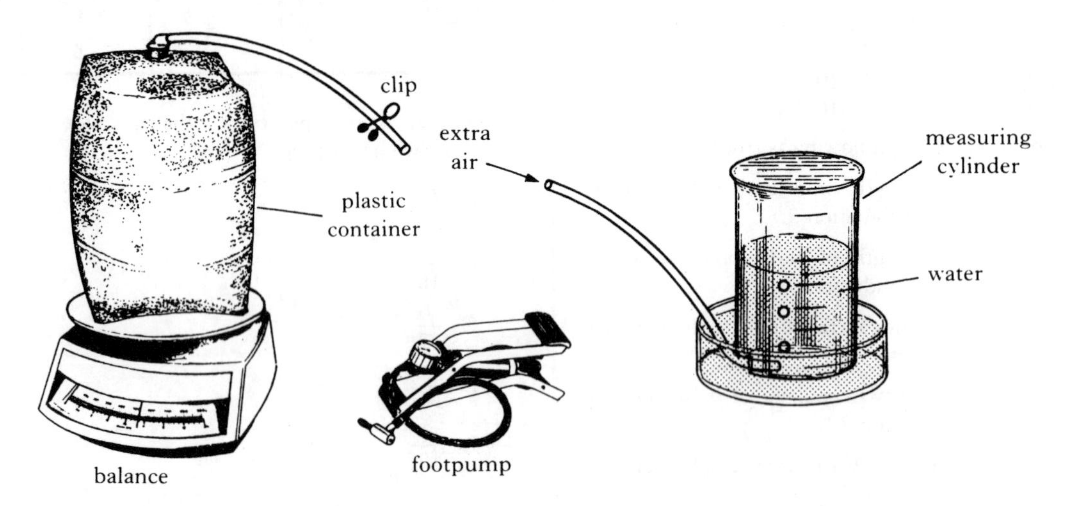

Using the footpump, extra air is pumped into the container. This extra air is released into the measuring cylinder as shown above and its volume measured.

The following measurements are recorded.

mass of container full of air = 362·00 g
mass of container with extra air = 363·86 g
volume of air released = 1687·00 cm^3

What value do these results give for the density of air in kg m^{-3}? **3**

Marks

33. Liquid nitrogen changes to its gaseous state at a temperature of −196 °C.

(*a*) What is this temperature in kelvin?

(*b*) Explain why a temperature of 0 kelvin is described as "the absolute zero of temperature". **3**

34. The circuit diagram for an **ideal** op–amp connected in the inverting mode is shown below.

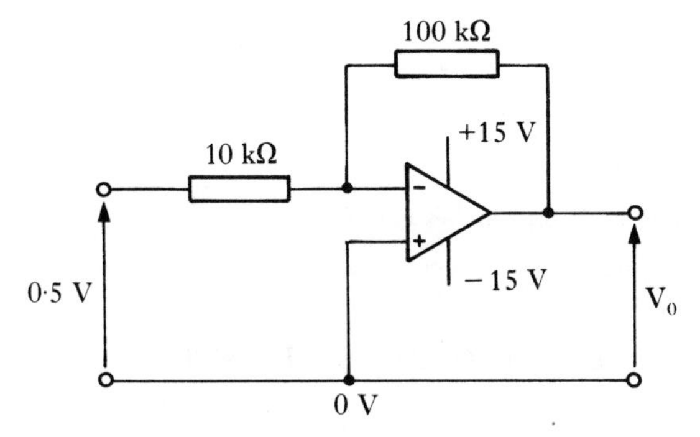

(*a*) What is the potential at the inverting input?

(*b*) What is the value of the current in the input resistor? **3**

35. Loudspeakers 1 and 2 are both connected to the same signal generator which is set to produce a 1 kHz signal.

Loudspeaker 1 is switched on but loudspeaker 2 is switched off.

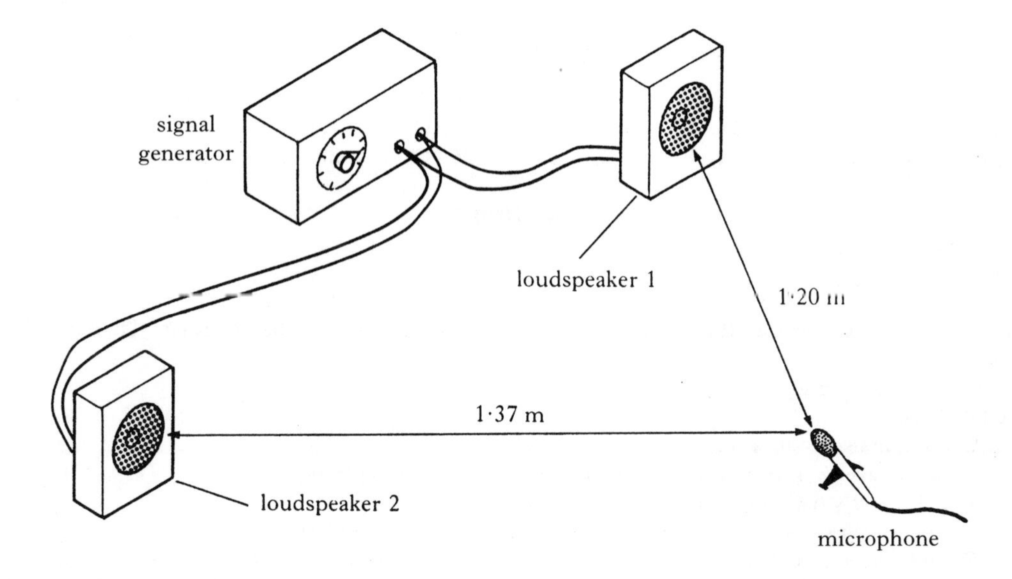

State **and** explain what happens to the amplitude of the signal picked up by the microphone when loudspeaker 2 is switched on.

Your explanation should include a calculation using the value of the speed of sound in air as 340 m s^{-1}. **3**

Marks

36. (*a*) Materials may be classified as "conductors", "semiconductors" and "insulators".

Give an example of a material from each of these groups.

(*b*) An electronics textbook states that

". . . n-type semiconductor material is formed by doping a pure semiconductor with impurity atoms."

What is meant by the term "n-type" semiconductor material? **3**

37. Energy is produced within the Sun by fusion reactions.

(*a*) State what is meant by a fusion reaction.

(*b*) Explain briefly why a fusion reaction releases energy. **3**

[*END OF QUESTION PAPER*]

HIGHER PHYSICS ANSWERS

Paper I

Section A

1. D	**2.** B	**3.** C	**4.** C	**5.** C	**6.** A	**7.** D	**8.** D	**9.** D	**10.**
11. D	**12.** C	**13.** C	**14.** B	**15.** D	**16.** C	**17.** A	**18.** B	**19.** A	**20.**
21. D	**22.** A	**23.** B	**24.** E	**25.** B	**26.** C	**27.** A	**28.** D	**29.** D	**30.**

Section B

31. Proof

32. $1{\cdot}10\ \text{kg m}^{-3}$

33. (*a*) +77 K

(*b*) Kinetic energy of particles theoretically zero so no more energy can be removed. Therefore lowest possi temperature.

34. (*a*) 0 volts (*b*) 50 μA

35. Amplitude will fall.
Wavelength of sound used = 0·34 m.
Sound from loudspeaker **2** travels further to the microphone than sound from **1**.
Path difference is exactly half of wavelength, hence destructive interference.

36. (*a*) Conductor — copper; semi-conductor — germanium; insulator — plastic.

(*b*) Semi-conductor material in which the majority of charge carriers are negative (i.e. electrons).

37. (*a*) Fusion reactions are those in which lighter nuclei fuse or join together to form a heavier nucleus and in the proc release energy.

(*b*) Mass is "lost" during the reaction and is transformed into energy, according to the relationship $E = mc^2$.

SCOTTISH CERTIFICATE OF EDUCATION
Part 1994

FRIDAY, 13 MAY
1.30 PM – 4.00 PM

PHYSICS
HIGHER GRADE
Paper II

Read carefully

1 All questions should be attempted.

2 Enter the question number clearly in the margin beside each question.

3 Any necessary data will be found in the Data Sheet.

4 Care should be taken not to give an unreasonable number of significant figures in the final answers to calculations.

5 Square-ruled paper (if used) should be placed inside the front cover of the answer book for return to the Scottish Qualifications Authority.

Marks

1. (*a*) A long jumper devises a method for estimating the horizontal component of his velocity during a jump.

His method involves first finding out how high he can jump **vertically**.

He finds that the maximum height he can jump is 0·86 m.

(i) Show that his initial vertical velocity is 4·1 m s^{-1}.

He now assumes that when he is long jumping, the initial vertical component of his velocity at take-off is 4·1 m s^{-1}.

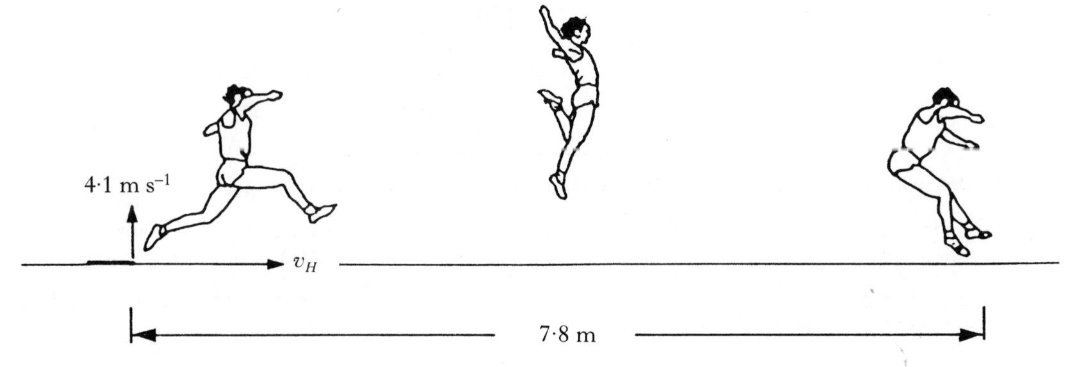

The length of his long jump is 7·8 m.

(ii) Calculate the value that he should obtain for the horizontal component of his velocity, v_H. **5**

(*b*) His coach tells him that, during the 7·8 m jump, his maximum height above the ground was less than 0·86 m. Ignoring air resistance, state whether his actual horizontal component of velocity was greater or less than the value calculated in part (*a*) (ii). You must justify your answer. **2**

(7)

Marks

2. (*a*) A hot air balloon, of total mass 500 kg, is held stationary by a single vertical rope.

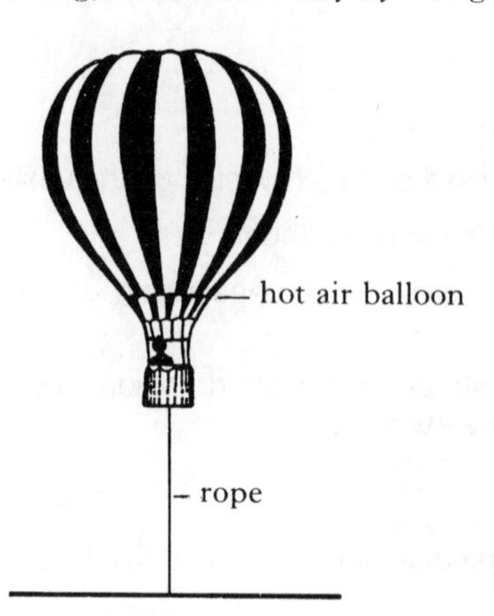

(i) Draw a sketch of the balloon. On your sketch, mark and label all the forces acting on the balloon.

(ii) When the rope is released, the balloon initially accelerates vertically upwards at $1{\cdot}5\ \mathrm{m\,s^{-2}}$. Find the magnitude of the buoyancy force.

(iii) Calculate the tension in the rope **before** it is released. **5**

(*b*) An identical balloon is moored using two ropes, each of which makes an angle of 25° to the vertical, as shown below.

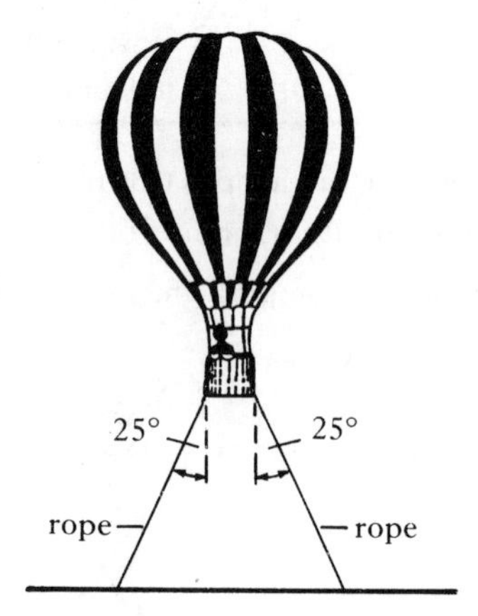

By using a scale diagram, or otherwise, calculate the tension in each rope. **2**

(*c*) During a flight, when a hot air balloon is travelling vertically upwards with constant velocity, some hot air is released. This allows cooler air to enter through the bottom of the balloon.

Describe **and** explain the effect of this on the motion of the balloon. You may assume that the volume of the balloon does not change. **3**

(10)

Marks

3. A water rocket consists of a plastic bottle, partly filled with water. Air is pumped in through the water as shown in Figure 1. When the pressure inside the bottle is sufficiently high, water is forced out at the nozzle and the rocket accelerates vertically upwards as shown in Figure 2.

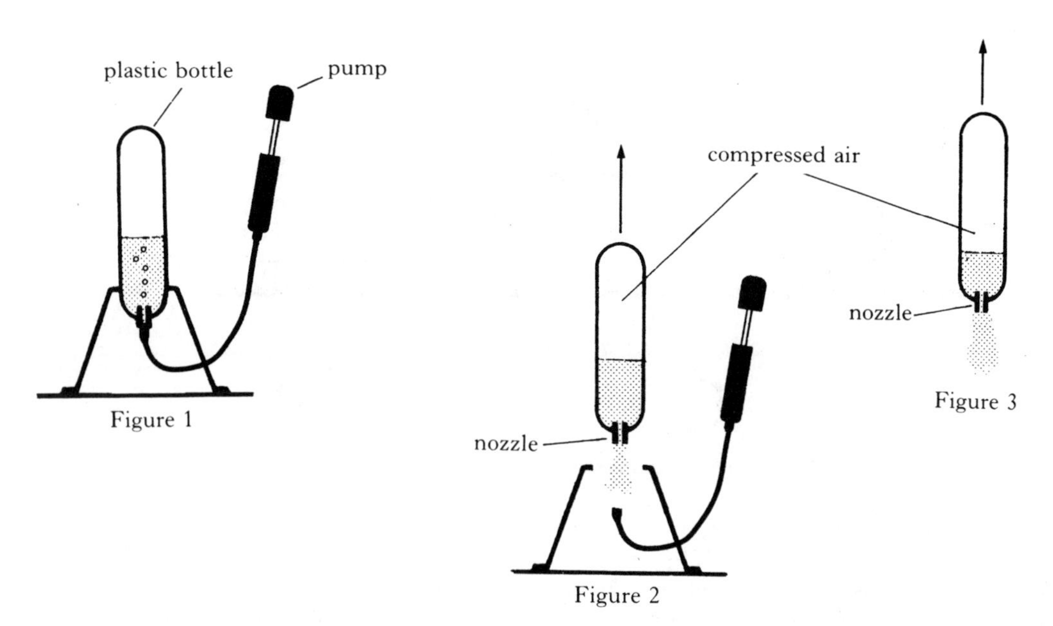

Figure 1

Figure 2

Figure 3

(*a*) (i) At take-off, the volume of air in the bottle is 750 cm^3 at a pressure of $1{\cdot}76 \times 10^5$ Pa.

Figure 3 shows the rocket at a later stage in its flight, when the volume of the air in the bottle has increased to 900 cm^3.

Calculate the new pressure of the compressed air at this later stage in its flight.

(ii) The area of the water surface which is in contact with the compressed air in the bottle is $5{\cdot}0 \times 10^{-3}$ m^2.

Calculate the force exerted on the water by the compressed air at the new pressure. **4**

(*b*) Explain fully why the rocket rises as the water is forced out at the nozzle. **2**

(6)

Marks

4. A student uses a linear air track to investigate collisions. In one experiment a vehicle, mass 0·50 kg, moves along and rebounds from a metal spring mounted at one end of the level track as shown below.

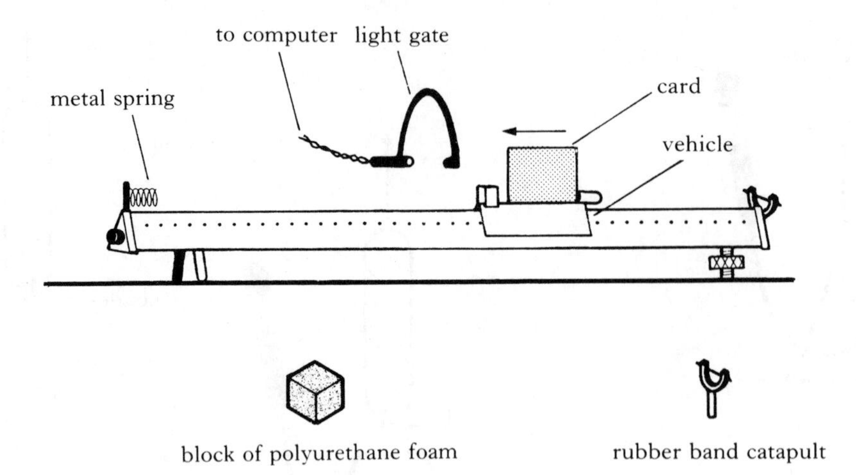

By using a light gate connected to a computer, she obtains values for the speed of the vehicle before and after it collides with the spring.

She then repeats this procedure, replacing the metal spring first with the block of polyurethane foam and then with the rubber band catapult. She records the results of each experiment in a table as shown below.

	Metal spring	Polyurethane block	Rubber band
Speed before collision/m s^{-1}	0·55	0·55	0·55
Speed after collision/m s^{-1}	0·49	0·33	0·43
Kinetic energy before collision/J	0·076	0·076	0·076
Kinetic energy after collision/J	0·060		

(*a*) Calculate values of kinetic energy to complete the last row of the table. **2**

(*b*) For which experiment is the collision most nearly elastic? You must justify your answer. **1**

(*c*) Describe a method she could use to give the vehicle the same initial speed each time. **1**

Marks

(*d*) In order to analyse a collision in more detail, she now uses a motion sensor. This enables the computer to display a velocity–time graph of the motion.

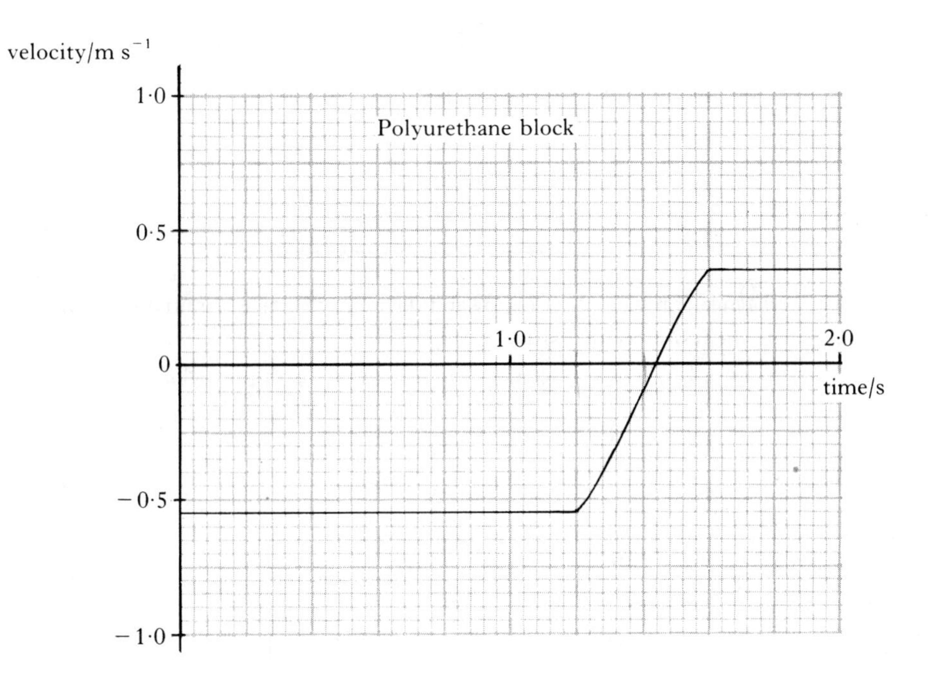

(i) Use information from this graph to calculate the average force exerted by the polyurethane block on the vehicle, mass 0·50 kg, during the time that they are in contact.

(ii) Describe the motion of the vehicle during the time that it is in contact with the polyurethane block. **4**

(8)

Marks

5. (*a*) A potential divider is used to provide an output voltage V_0 from a 10 V supply as shown below. The supply has negligible internal resistance.

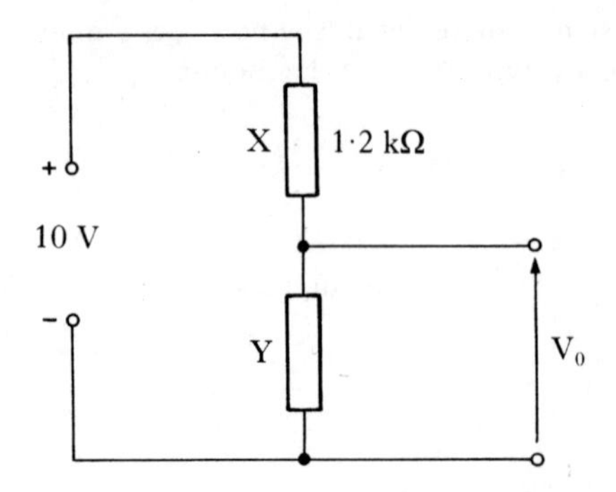

(i) The resistance of resistor X is 1·2 kΩ and the output voltage required is 6·0 V. Calculate the resistance of resistor Y.

(ii) A load resistor Z is now connected across the output as shown below.

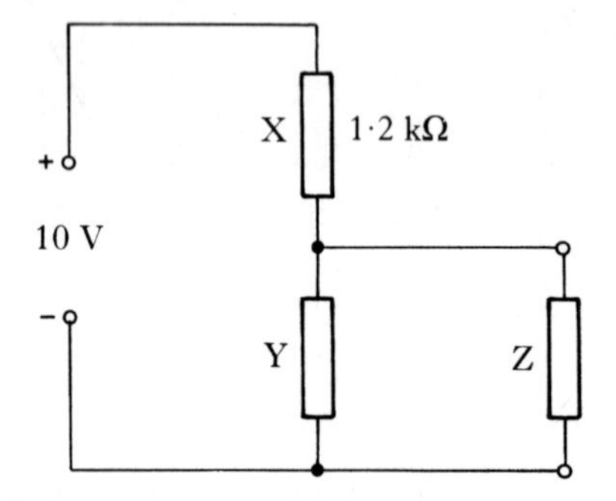

Explain why the voltage across Z is less than 6·0 V.

(iii) Calculate the voltage across resistor Z when its resistance is 4·7 kΩ. **6**

(*b*) A Wheatstone bridge circuit is shown below.

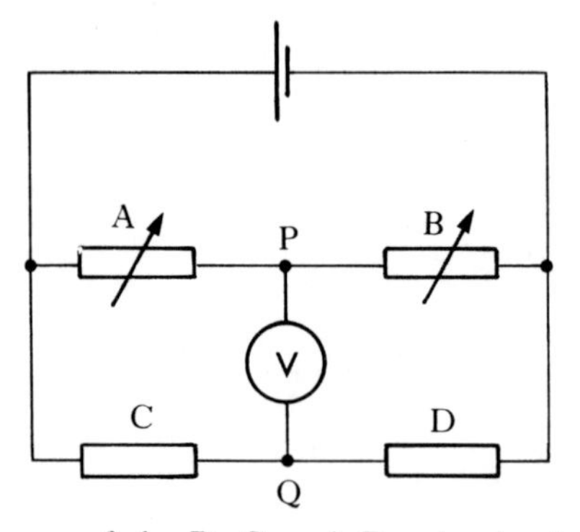

(i) How are the resistances of A, B, C and D related when the bridge is balanced?

Marks

(ii) C and D are fixed resistors, each of value 120 Ω. The resistors A and B are variable and each is initially set at 120 Ω. The voltmeter is used to measure the p.d. between the points P and Q.

Small changes are made to the resistances of A and B, and the various values are shown in the table below.

Resistance of A/Ω	Resistance of B/Ω	Voltmeter reading/mV
120	120	
121	120	− 21
121	121	
121	122	
121	119	

Copy and complete the last column of the table to show the voltmeter readings (including sign) that you would expect for each of the remaining sets of resistance values. **2**

(8)

Marks

6. (*a*) A rechargeable cell is rated at 0·50 A h (ampere hour). This means that, for example, it can supply a constant current of 0·50 A for a period of 1 hour. The cell then requires to be recharged.

(i) What charge, in coulombs, is available from a fully charged cell?

(ii) A fully charged cell is connected to a load resistor and left until the cell requires recharging. During this time, the p.d. across the terminals of the cell remains constant at 1·2 V.

Calculate the electrical energy supplied to the load resistor in this case. **3**

(*b*) (i) State what is meant by the e.m.f. of a cell.

(ii) The circuit shown below is used in an experiment to find the e.m.f. and internal resistance of the rechargeable cell.

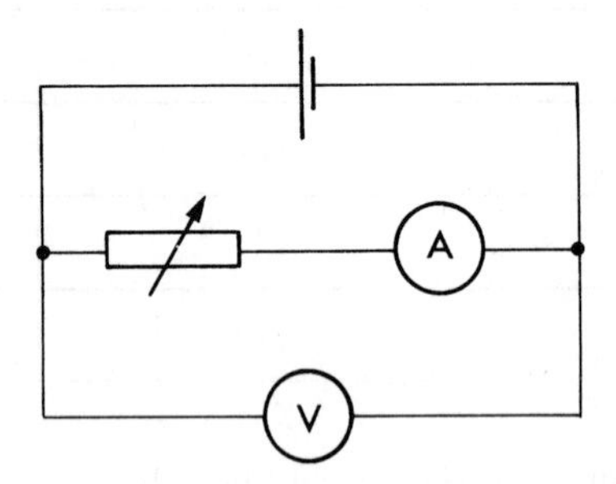

The voltmeter and ammeter readings for a range of settings of the variable resistor are used to produce the graph below.

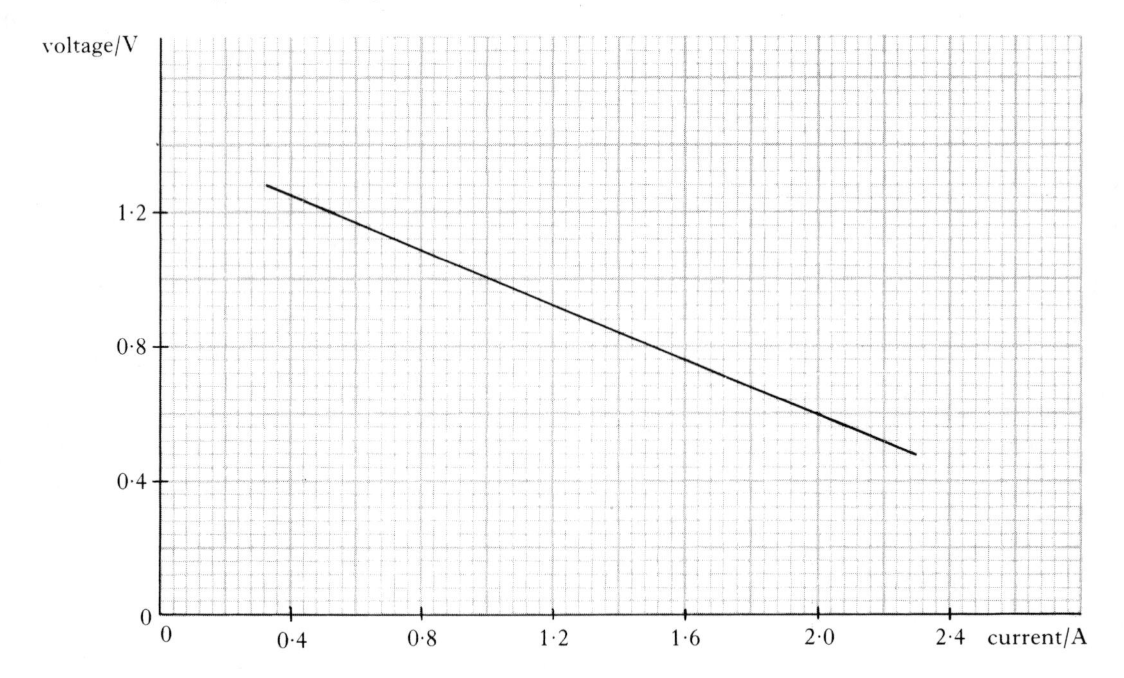

Use the graph to find the values for the e.m.f. **and** internal resistance of the cell. **4**

(7)

Marks

7. (*a*) In order to compare the brightness of a number of low voltage lamps, a solar cell is used to detect the light from the lamps. An operational amplifier, working in the inverting mode, is used to amplify the solar cell voltage.

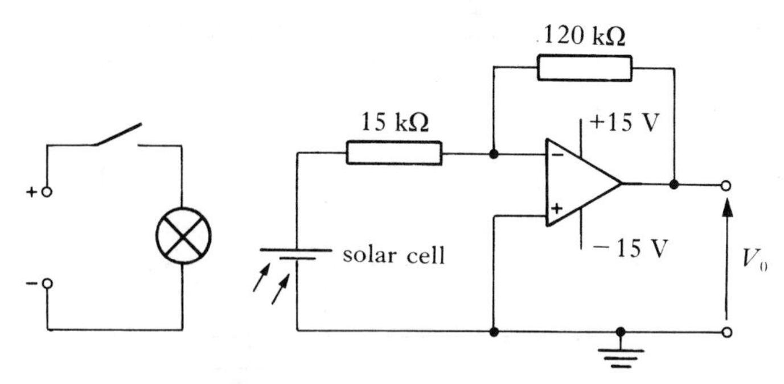

The apparatus is set up near to a window and, with the lamp switched off, there is an output voltage V_0 of $-1{\cdot}75$ V.

(i) Explain why the output voltage V_0 of the operational amplifier is not zero.

(ii) Calculate the solar cell voltage. **3**

(*b*) With the solar cell in the same position, the circuit is now altered so that the operational amplifier is working in the differential mode as shown below.

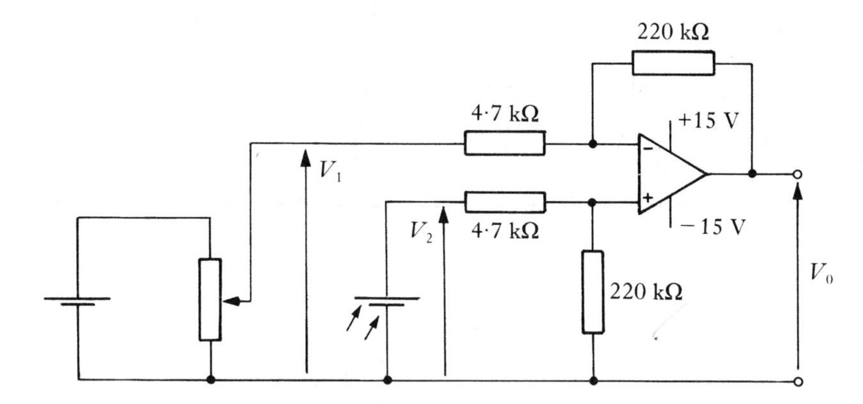

(i) With the lamp still unlit, the potentiometer setting is adjusted until the output voltage is zero.

Explain how this circuit enables the output voltage to be set to zero volts.

(ii) With V_1 unchanged, the lamp is switched on and the output voltage V_0 is now $1{\cdot}50$ V.

Calculate the voltage which the solar cell now produces. **4**

(7)

Marks

8. The circuit shown below is used to investigate the charging and discharging of a capacitor.

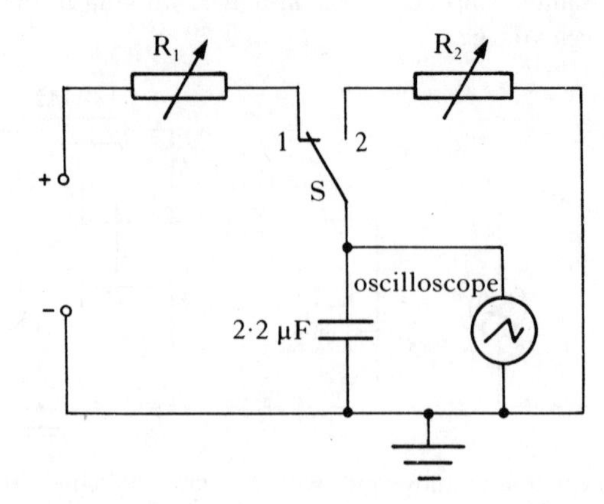

The capacitor is repeatedly charged and discharged by switching S between contacts 1 and 2.

(*a*) For one setting of R_1 and R_2 the following trace is obtained on the oscilloscope.

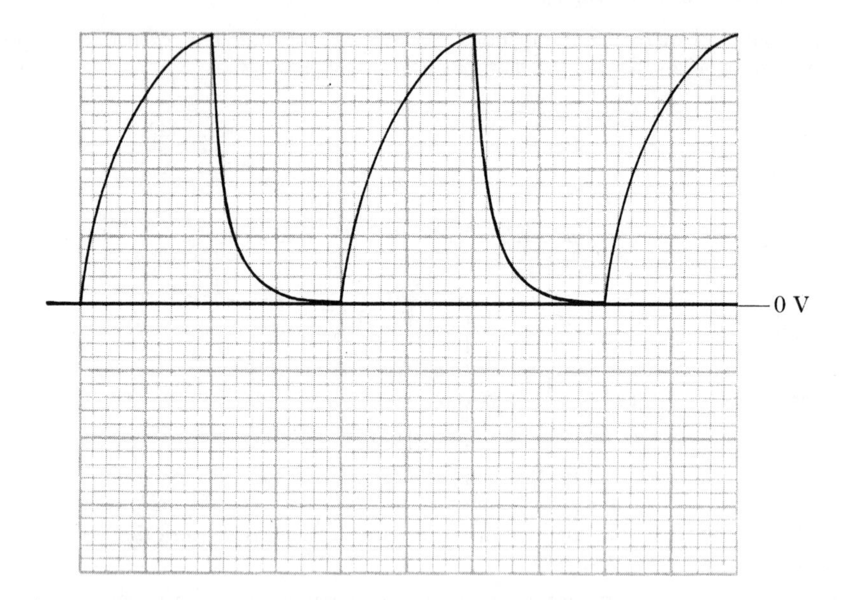

The time base of the oscilloscope is set at 5 ms per centimetre and the Y gain is set at 2 V per centimetre.

Calculate the frequency of the vibrating switch. **2**

Marks

(*b*) With the settings of the oscilloscope unaltered, a new trace is produced when the resistance of one of the variable resistors is changed.

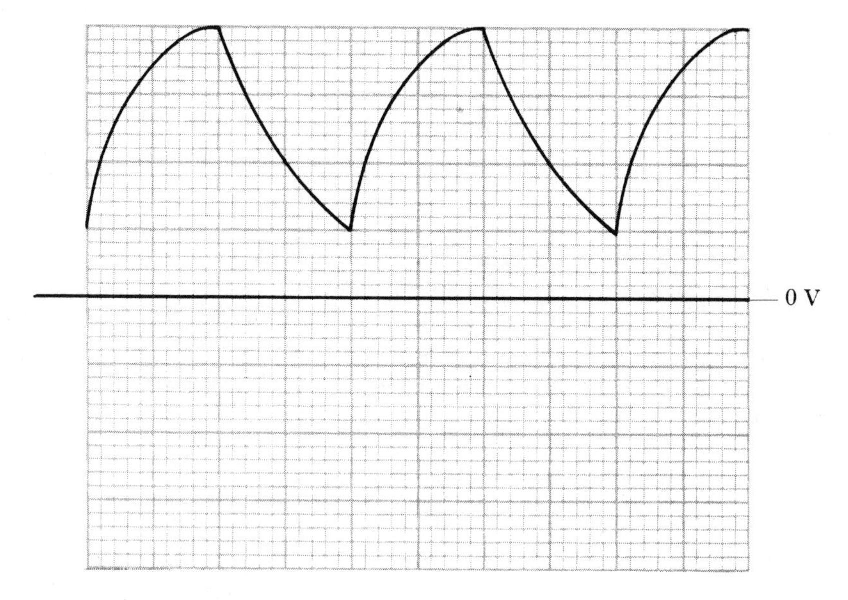

(i) State which one of the variable resistors was changed and whether its resistance was increased or decreased. Justify your answer.

(ii) Calculate the charge lost by the capacitor each time the switch is in the discharge position. **6**

(8)

Marks

9. (*a*) The diagram below shows the path of a monochromatic beam of light through a triangular plastic prism.

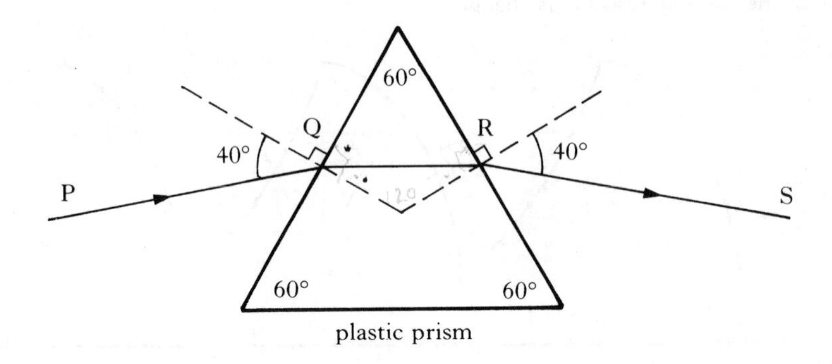

plastic prism

(i) Calculate the refractive index of the plastic.

(ii) Sketch a copy of the above diagram with ray PQRS clearly labelled. (Sizes of angles need not be shown.)

Add to your drawing the path which the ray PQ would take from Q if the prism were made of a plastic with a **slightly higher** refractive index. **3**

(*b*) The original prism is now replaced with one of the same size and shape but made from glass of refractive index 1·80.

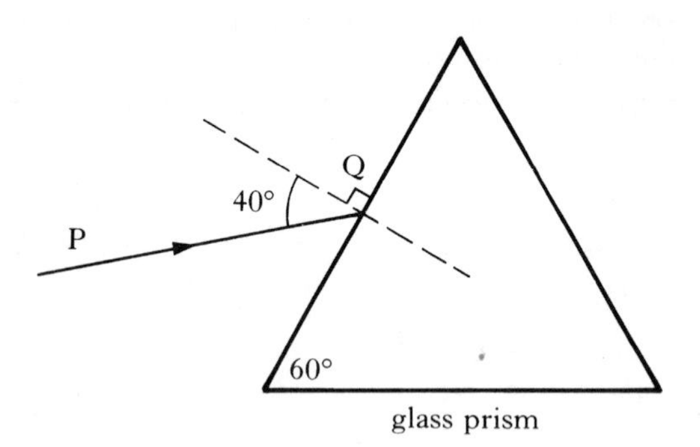

glass prism

(i) Calculate the critical angle for this glass.

(ii) Draw an accurate diagram, showing the passage of the ray PQ through this prism until after it emerges into the air.

Mark on your diagram the values of all relevant angles. **5**

(8)

SCOTTISH CERTIFICATE OF EDUCATION 1995

WEDNESDAY, 17 MAY
9.30 AM – 11.00 AM

PHYSICS
HIGHER GRADE
Paper I

Read Carefully

1 All questions should be attempted.

2 The following data should be used when required unless otherwise stated.

Speech of light in vacuum c	$3{\cdot}00 \times 10^{8}$ m s^{-1}	Planck's constant h	$6{\cdot}63 \times 10^{-34}$ J s
Charge on electron e	$-1{\cdot}60 \times 10^{-19}$ C	Mass of electron m_e	$9{\cdot}11 \times 10^{-31}$ kg
Acceleration due to gravity g	$9{\cdot}8$ m s^{-2}	Mass of proton m_p	$1{\cdot}67 \times 10^{-27}$ kg

Section A (questions 1 to 30)

3 Check that the answer sheet is for Physics Higher I (Section A).

4 Answer the questions numbered 1 to 30 on the answer sheet provided.

5 Fill in the details required on the answer sheet.

6 Rough working, if required, should be done only on this question paper, or on the first two pages of the answer book provided—**not** on the answer sheet.

7 For each of the questions 1 to 30 there is only **one** correct answer and each is worth 1 mark.

8 Instructions as to how to record your answers to questions 1–30 are previously given.

Section B (questions 31 to 38)

9 Answer questions numbered 31 to 38 in the answer book provided.

10 Fill in the details on the front of the answer book.

11 Enter the question number clearly in the margin of the answer book beside each of your answers to questions 31 to 38.

12 Care should be taken **not** to give an unreasonable number of significant figures in the final answers to calculations.

SECTION A

Answer questions 1–30 on the answer sheet.

1. Which one of the following is a vector quantity?

A Distance

B Time

C Speed

D Energy

E Weight

2. The velocity-time graph of the motion of an object starting from rest is shown below.

velocity/m s^{-1}

10
5
0
−5
1 2 3 4 5 6
time/s

Which of the following statements about the motion of the object is/are true?

I There is a change of direction of the object at 4 s.

II The acceleration-time graph is of the form shown below.

acceleration/m s^{-2}

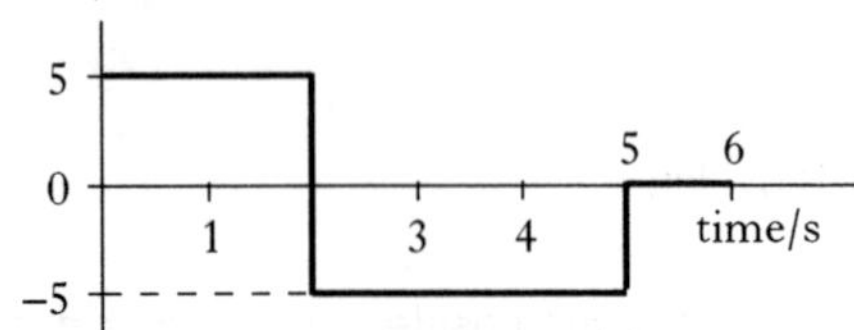

III The displacement of the object from the starting point is greatest at 6 s.

A I only

B II only

C I and II only

D I and III only

E II and III only

3. A cannonball is fired horizontally at 40 m s^{-1} from the top of a vertical cliff and it hits its target. The height of the cliff above the level of the sea is 80 m.

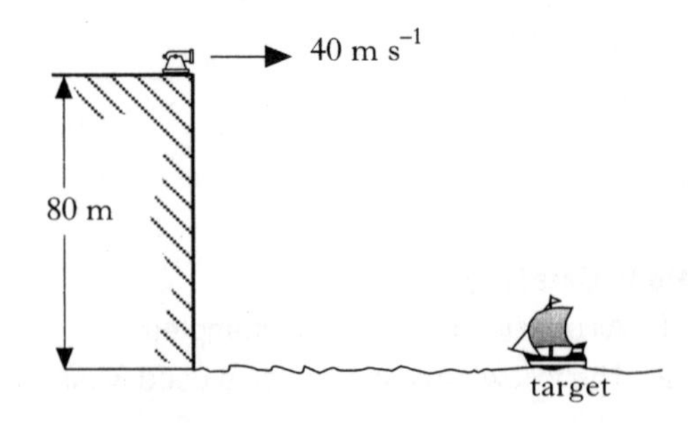

How far is the target from the foot of the cliff, if air resistance is negligible and the acceleration due to gravity is 10 m s^{-2}?

A 320 m

B 160 m

C 80 m

D 45 m

E 40 m

4. A golfer strikes a golf ball which then moves off at an angle to the ground. The ball, following the path shown below, lands 6 s later.

The graphs below show how the ball's horizontal and vertical components of velocity vary with time, the acceleration due to gravity being $10\,m\,s^{-2}$.

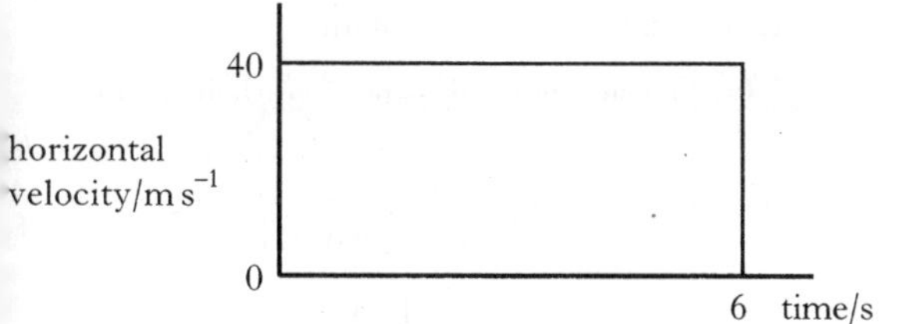

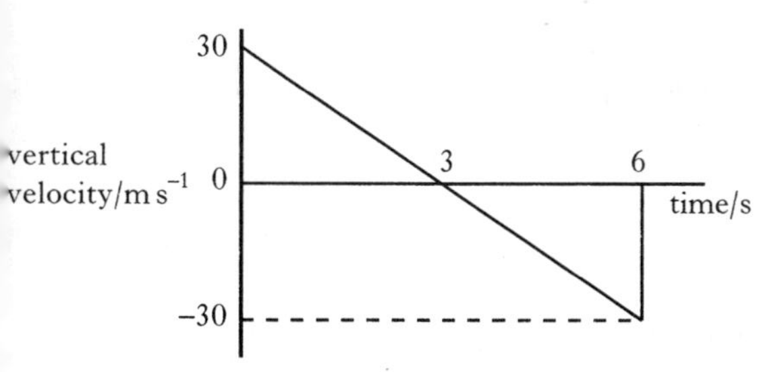

What is the speed of the ball just before it hits the ground?

A $10\,m\,s^{-1}$

B $30\,m\,s^{-1}$

C $40\,m\,s^{-1}$

D $50\,m\,s^{-1}$

E $70\,m\,s^{-1}$

5. A tension force of 180 N is applied vertically upwards to a box of mass 15 kg.

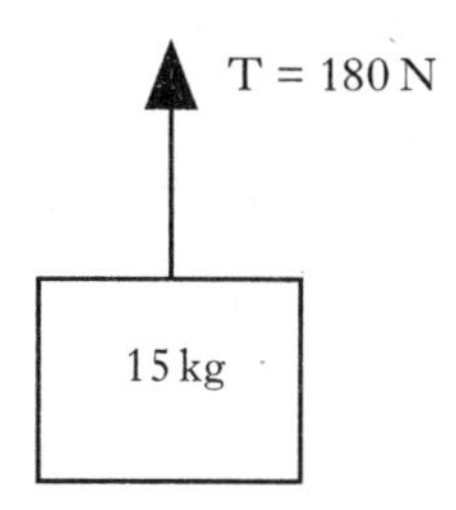

Assuming that the acceleration due to gravity is $10\,m\,s^{-2}$, the acceleration of the box is

A $2\,m\,s^{-2}$

B $8\,m\,s^{-2}$

C $10\,m\,s^{-2}$

D $12\,m\,s^{-2}$

E $20\,m\,s^{-2}$.

6. Two boxes on a frictionless horizontal surface are joined together by a string, as shown.

The 4 kg box is being pulled to the right by a constant horizontal force of 12 N.

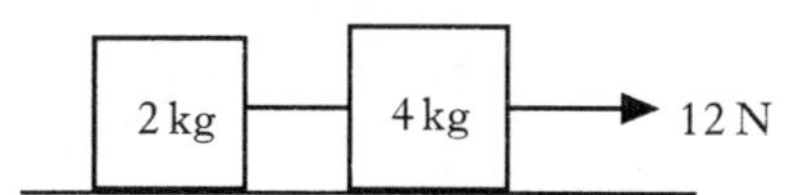

What is the value of the force of tension in the string joining the two boxes?

A 2 N

B 4 N

C 6 N

D 8 N

E 12 N

7. The total mass of a motorcycle and rider is 250 kg. During braking, they are brought to rest from a speed of $15\,m\,s^{-1}$ in a time of 10 s. The maximum energy which could be converted to heat by the brakes is

A 3 750 J

B 28 125 J

C 37 500 J

D 56 250 J

E 375 000 J.

8. A model car of mass 3 kg, initially at rest, is acted upon by an unbalanced force F, as shown in the following force-time graph.

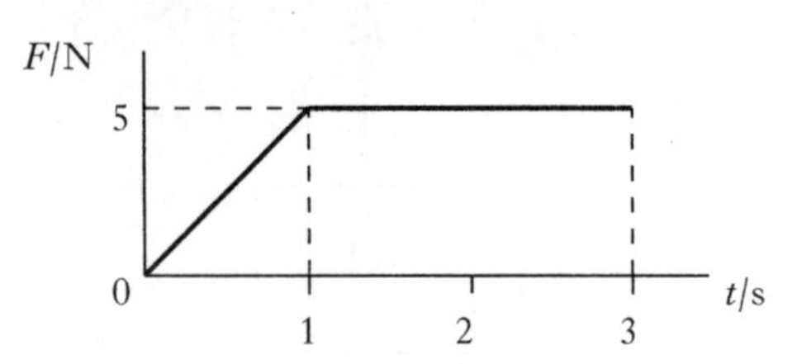

What is the momentum of the model car at time $t = 3$ s?

A $0\,\text{kg m s}^{-1}$

B $2{\cdot}5\,\text{kg m s}^{-1}$

C $5\,\text{kg m s}^{-1}$

D $12{\cdot}5\,\text{kg m s}^{-1}$

E $15\,\text{kg m s}^{-1}$

9. A rectangular box of mass 10 kg is lying on a flat surface on a planet where the gravitational field strength is $4\,\text{N kg}^{-1}$.

The base of the box measures 4 m by 2 m.

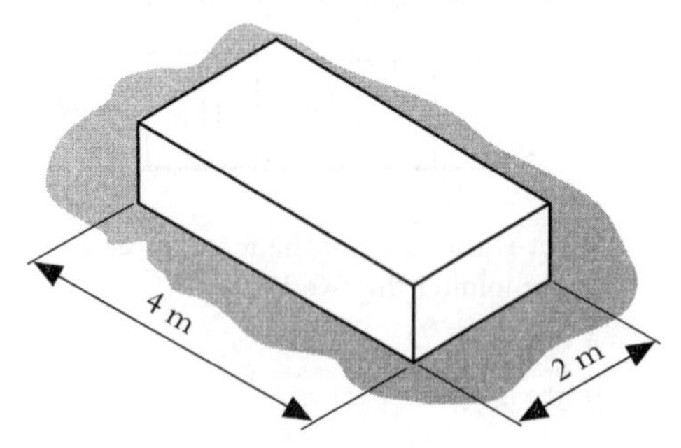

Which of the following statements is/are correct?

I The weight of the box is 100 N.

II The weight of the box is 40 N.

III The pressure which the box exerts on the flat surface is 5 Pa.

A I only

B II only

C III only

D I and III only

E II and III only

10. The end of a bicycle pump is sealed with a small rubber stopper. The air in chamber C is now trapped.

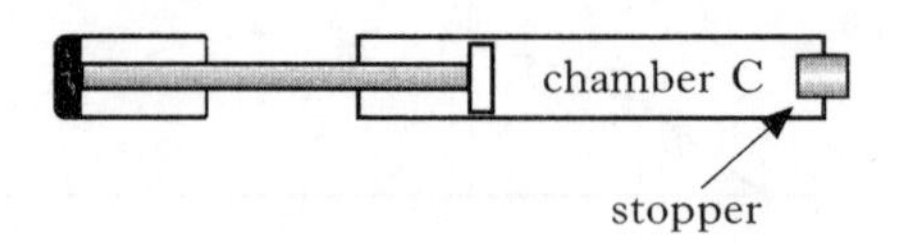

The plunger is then pushed in slowly, causing the air in the chamber C to be compressed. As a result of this, the pressure of the air increases.

Which of the following explain(s) why the pressure increases, assuming that the temperature remains constant?

I The air molecules increase their average speed.

II The air molecules are colliding more often with the walls of the chamber.

III Each air molecule is striking the walls of the chamber with greater force.

A II only

B III only

C I and II only

D I and III only

E I, II and III

11. An electron is accelerated from rest in an electron gun, across a potential difference of 2×10^3 V.

The kinetic energy gained by the electron as it goes through the electron gun is

A $8{\cdot}0 \times 10^{-23}$ J

B $8{\cdot}0 \times 10^{-20}$ J

C $3{\cdot}2 \times 10^{-19}$ J

D $1{\cdot}6 \times 10^{-16}$ J

E $3{\cdot}2 \times 10^{-16}$ J.

12. A student requires a resistor for an electronics project and its value must lie in the range $(15 \pm 3)\ \Omega$.

The only resistors available have values of exactly $10\ \Omega$.

Which of the following combinations of these $10\ \Omega$ resistors could be used?

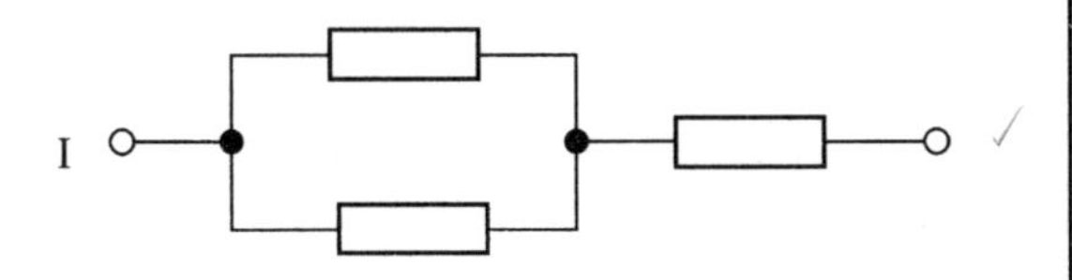

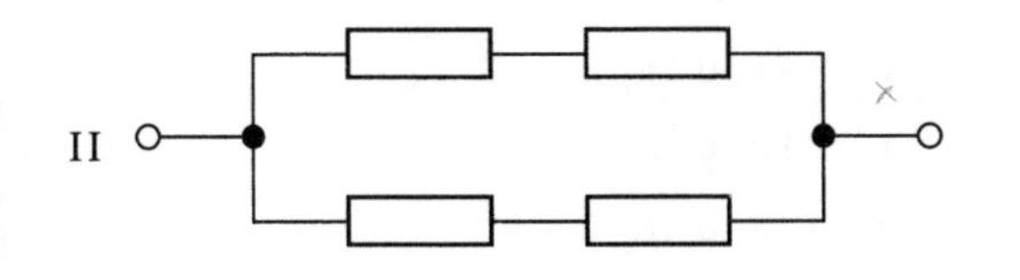

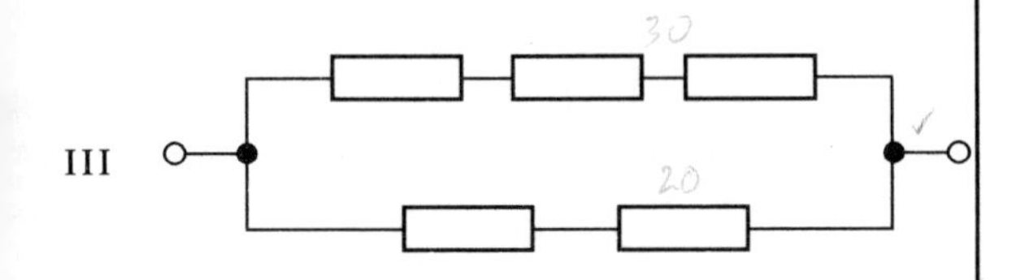

A I only

B I and II only

C I and III only

D II and III only

E I, II and III

13. In the following circuit, what is the potential difference across the $12\ \Omega$ resistor when the switch S is (i) open, and (ii) closed? The supply has negligible internal resistance.

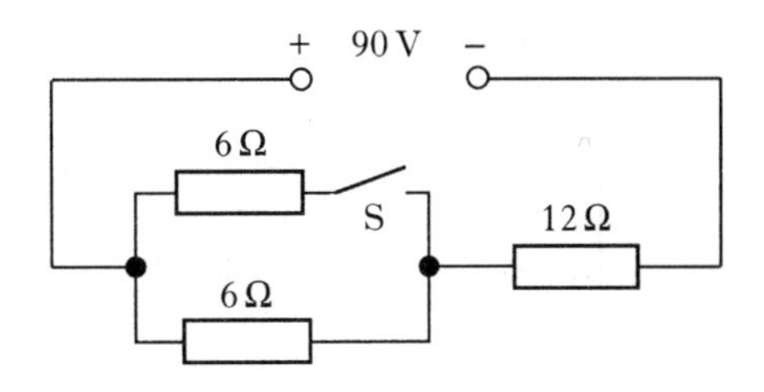

	(i) p.d. when switch S open	*(ii) p.d. when switch S closed*
A	30 V	18 V
B	45 V	45 V
C	60 V	45 V
D	60 V	72 V
E	72 V	60 V

14. The circuit below shows two resistors connected to a 6 V d.c. supply of negligible internal resistance.

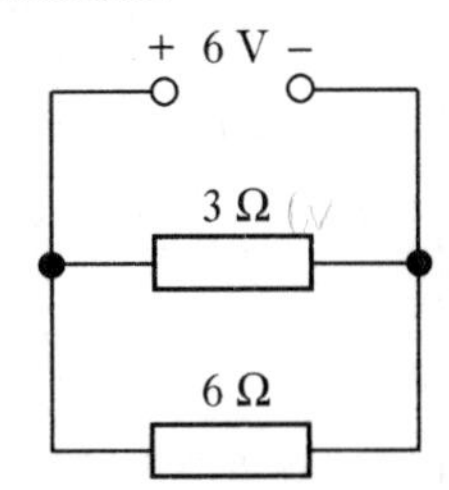

The power dissipated in the $3\ \Omega$ resistor is

A 3 W

B 6 W

C 9 W

D 12 W

E 18 W

15. An alternating voltage signal is displayed on an oscilloscope, with the settings shown.

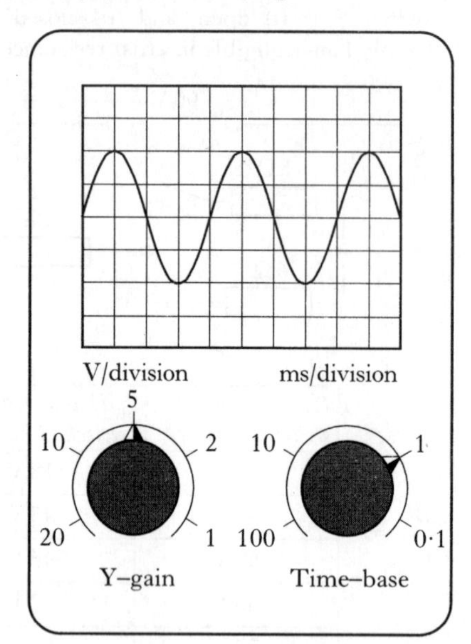

Which row in the following table gives the correct values for the peak voltage and frequency of the signal?

	Peak Voltage/V	*Frequency*/Hz
A	10	100
B	10	250
C	20	250
D	10	500
E	20	1000

16. The heating element in a boiler operates at 2400 W from a 120 V r.m.s. power supply.

What is the r.m.s. current, in amperes, in this element?

A 10

B $\frac{20}{\sqrt{2}}$

C 20

D $20\sqrt{2}$

E 40

17. The "coulomb per volt" is a unit of

A charge

B energy

C power

D capacitance

E potential difference.

18. The energy stored in a 500 μF capacitor charged to a voltage of 20 V is

A 5×10^{-3} J

B $2{\cdot}5\times10^{-2}$ J

C 5×10^{-2} J

D 1×10^{-1} J

E 2×10^{-1} J.

19. In the following circuit, a capacitor is being charged up from a d.c. source of e.m.f. E. The capacitor has a resistor in series with it, as shown.

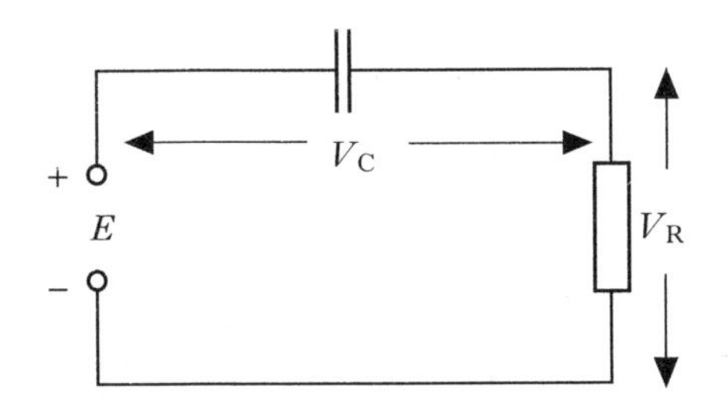

The voltages, V_C and V_R, across the components are recorded at regular time intervals as the capacitor charges up.

Which of the pairs of graphs shown below correctly represents the voltages across the capacitor and the resistor during charging?

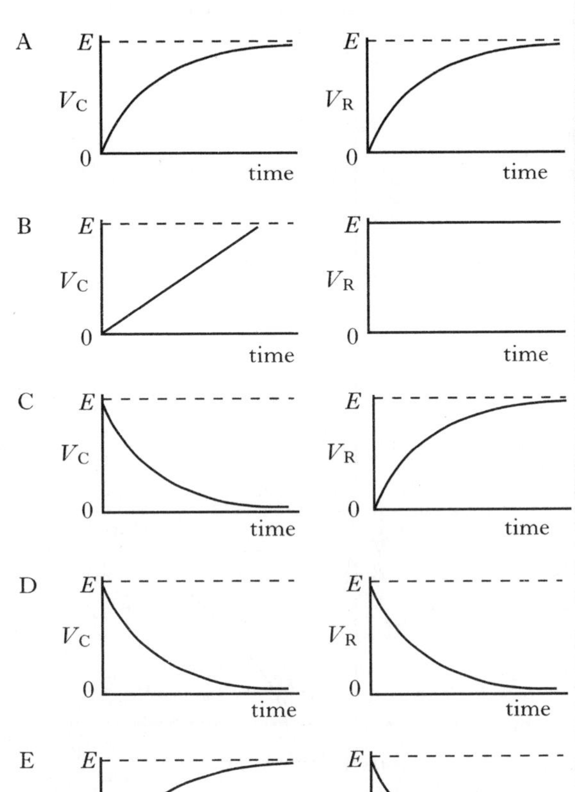

20. An oscilloscope is used to measure the frequency of the output voltage from an operational amplifier.

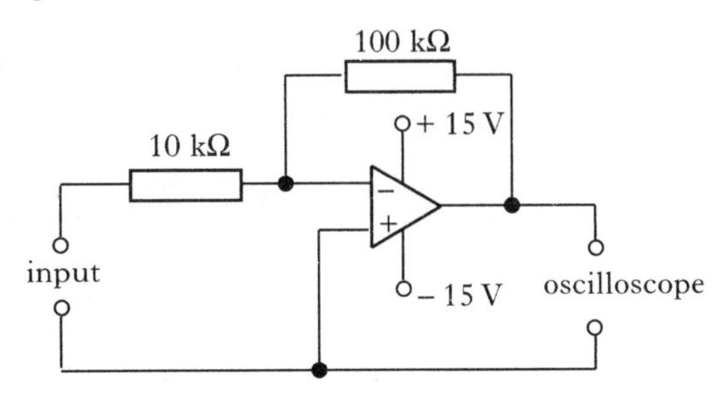

The input voltage has a frequency of 280 Hz and a peak value of 0·5 V.

The frequency of the output voltage is

A 14 Hz

B 28 Hz

C 280 Hz

D 560 Hz

E 2800 Hz.

21. A ray of light passing from air into water is refracted towards the normal.

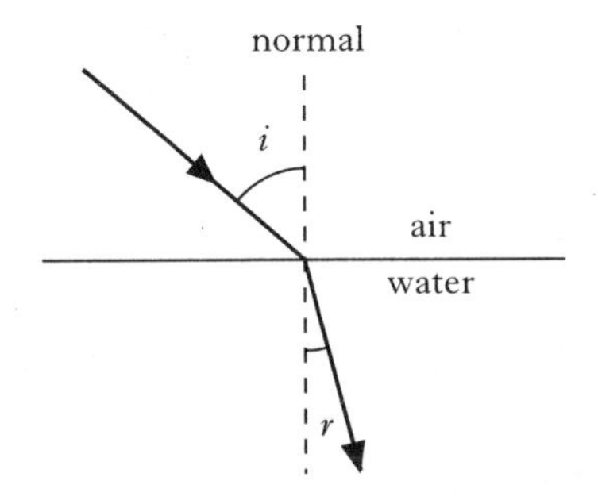

Which of the following statements is/are true?

I The speed of the light in water is less than the speed of the light in air.

II The frequency of the light in water is less than the frequency of the light in air.

III The wavelength of the light in water is less than the wavelength of the light in air.

A I only

B III only

C I and II only

D I and III only

E I, II and III

22. A ray of monochromatic light is directed towards a glass prism and travels through it.

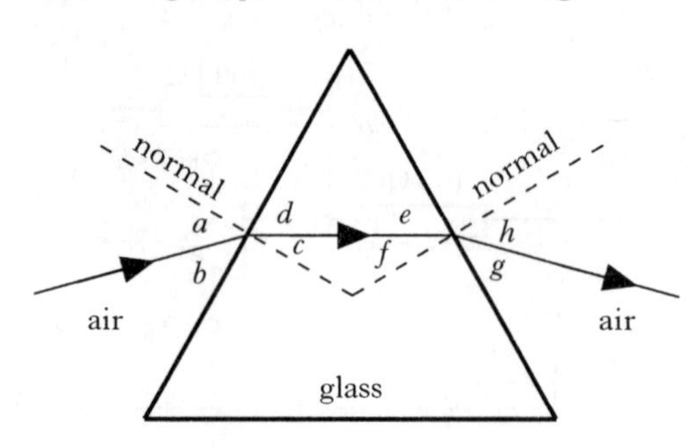

Which of the following expressions can be used to calculate the refractive index of the glass used for this prism?

A $\frac{\sin c}{\sin a}$

B $\frac{\sin b}{\sin c}$

C $\frac{\sin f}{\sin h}$

D $\frac{\sin h}{\sin f}$

E $\frac{\sin e}{\sin h}$

23. The energy, E, of a photon of light depends on its wavelength λ.

Which of the following graphs correctly illustrates the relationship between E and λ?

A

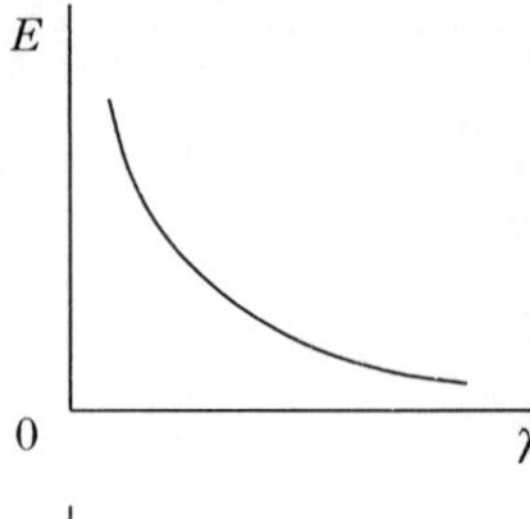

B

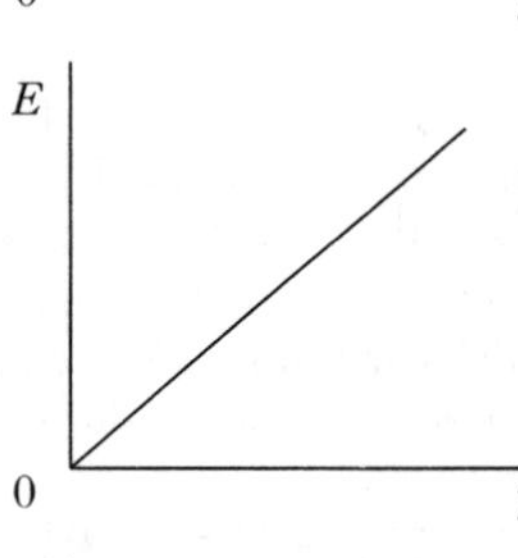

C

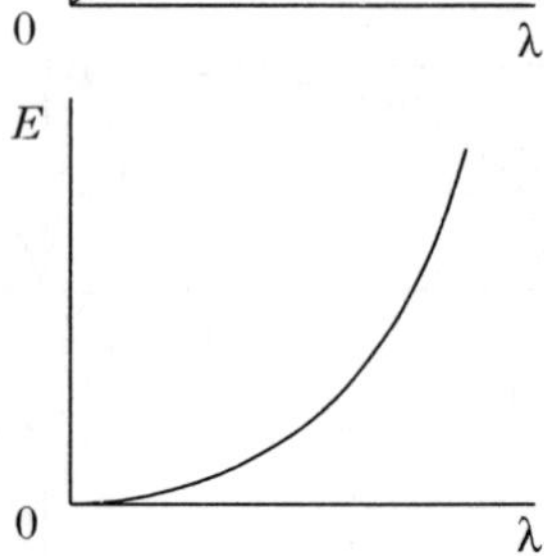

D

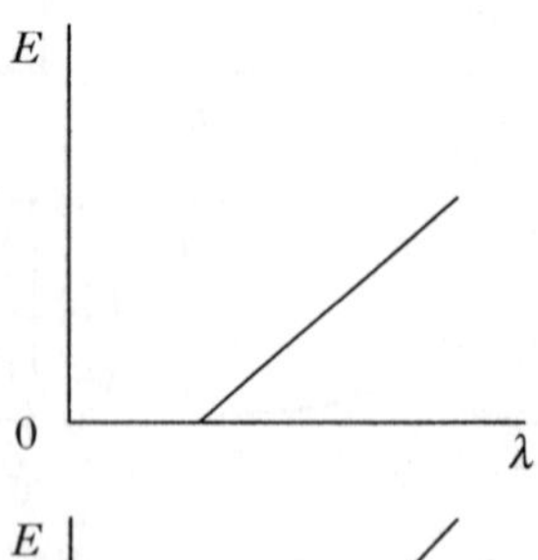

E

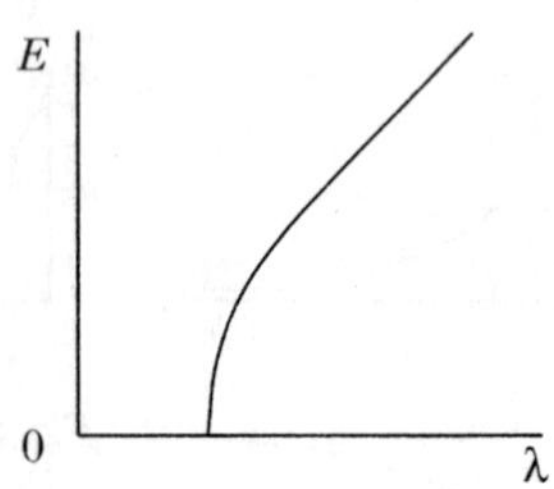

24. Two loudspeakers LS1 and LS2, connected to the same output of a signal generator, provide coherent sources of sound waves. A microphone, connected to an oscilloscope, is used to detect the sound.

Position P is the same distance from LS1 as it is from LS2. Position Q is **one wavelength** of the sound wave further from LS1 than it is from LS2.

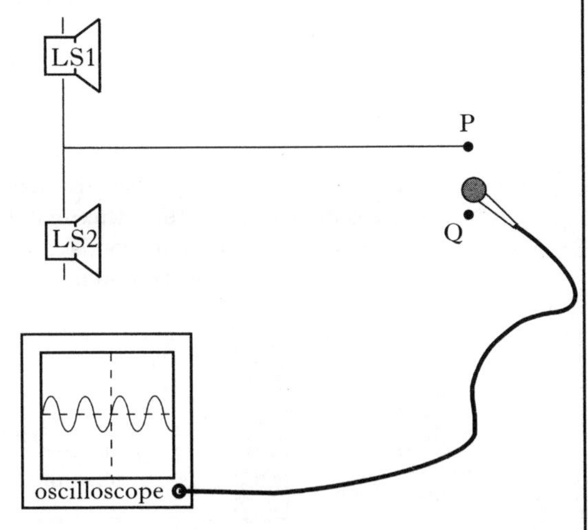

Which of the following best describes what happens to the oscilloscope trace as the microphone is slowly moved from position P to position Q?

A Constant amplitude of trace when moved from P to Q

B Minimum amplitude at P increasing to maximum amplitude at Q

C Maximum amplitude at P decreasing to minimum amplitude at Q

D Minimum amplitude at P, going through a maximum and then back to a minimum amplitude at Q

E Maximum amplitude at P, going through a minimum and then back to a maximum amplitude at Q

25. When monochromatic light is passed through a grating, a pattern of maxima and minima is observed as shown below.

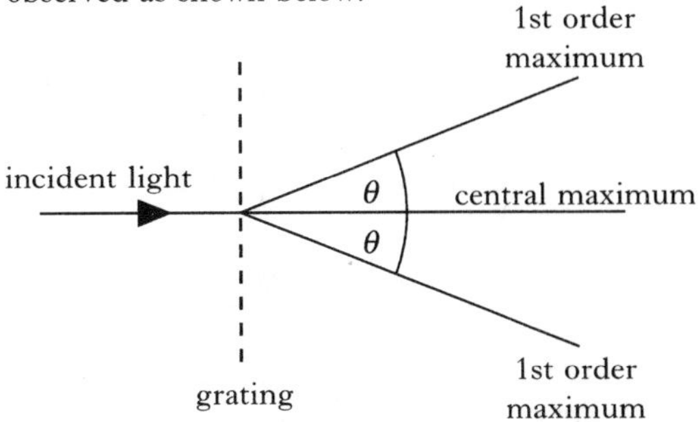

Which row in the following table represents the arrangement which would produce the greatest angle θ between the central and first order maxima?

	Grating (lines per mm)	*Colour of light*
A	100	Red
B	100	Green
C	100	Blue
D	200	Red
E	200	Blue

26. A point source S emits radiation equally in all directions.

Source

S P Q

The distance from S to Q is nine times the distance from S to P.

The intensity of radiation at P is I. The intensity at point Q is

A $9I$

B $3I$

C $\frac{I}{3}$

D $\frac{I}{9}$

E $\frac{I}{81}$.

27. In a laser, a photon of light is emitted when an electron makes a transition from a higher energy level to a lower one, as shown below.

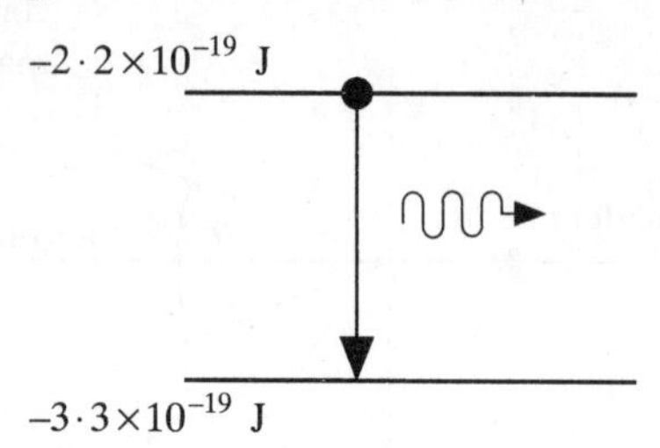

If the energy in each pulse of light from the laser is 10 J, how many photons are there in each pulse?

A $\dfrac{10}{5 \cdot 5 \times 10^{-19}}$

B $\dfrac{10}{(1 \cdot 1 + 1 \cdot 6) \times 10^{-19}}$

C $\dfrac{10}{3 \cdot 3 \times 10^{-19}}$

D $\dfrac{10}{2 \cdot 2 \times 10^{-19}}$

E $\dfrac{10}{1 \cdot 1 \times 10^{-19}}$

28. An element X emits an alpha particle to form a new element.

Which of the following statements is/are correct about this **new** element?

I The total number of protons and neutrons is 4 less than in element X.

II The number of protons is the same as in element X.

III The new element is an isotope of element X.

A I only

B II only

C III only

D I and III only

E II and III only

29. Which row in the following table shows the correct units for all three quantities listed?

	Absorbed Dose	*Dose Equivalent*	*Activity*
A	gray	sievert	becquerel ✓
B	becquerel	gray	sievert
C	sievert	becquerel	gray
D	sievert	gray	becquerel
E	gray	becquerel	sievert

30. A 60 mm thick lead absorber is placed between a gamma source and a detector. The reading measured by the detector is 240 Bq. The half-value thickness of the lead is 30 mm.

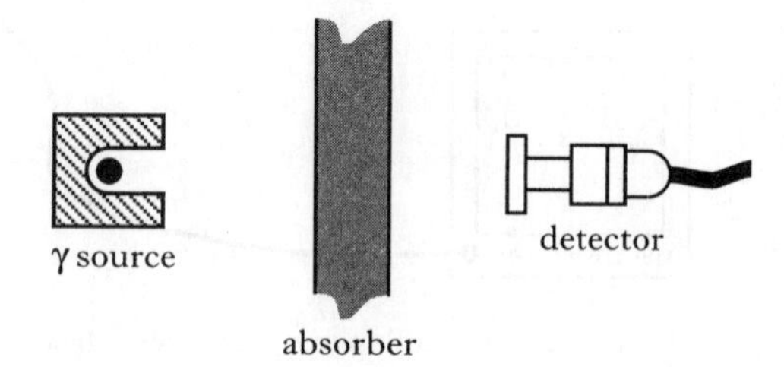

What will the reading be if the 60 mm absorber is replaced by one of thickness 120 mm?

A 120 Bq

B 80 Bq

C 60 Bq

D 40 Bq

E 30 Bq

SECTION B

Write your answers to questions 31 to 38 in the answer book.

Marks

31. An advertising brochure for a car gives the information that the car, starting from rest, can cover 400 m in 17·5 s under constant acceleration.

Calculate the acceleration of the car. **2**

32. A trolley of mass 2·0 kg rolls down a slope which makes an angle of 30° with the horizontal.

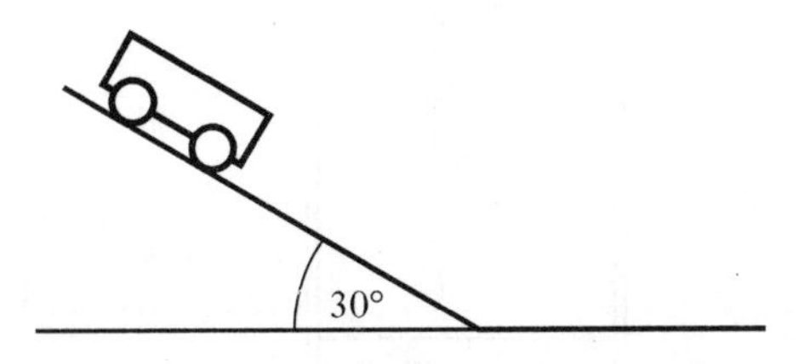

The constant frictional force opposing the motion is 4·0 N.

Calculate the size of the resultant force, in newtons, acting on the trolley. **2**

33. A skin diver carries her air supply in a steel cylinder on her back. She works at a depth where the pressure is $2{\cdot}5 \times 10^5$ Pa. When full, the cylinder contains $0{\cdot}060\,\text{m}^3$ of air at a pressure of $1{\cdot}6 \times 10^7$ Pa.

Calculate the volume of air available to her at this depth from a full cylinder. **3**

34. The potential difference across a lamp is 12 V.

How much energy is dissipated in the lamp when a charge of 5 C passes through it? **2**

35. A resistor and capacitor are connected in series with an a.c. supply of voltage V_S as shown below. A voltage V_C is produced across the capacitor. The frequency of the supply can be changed.

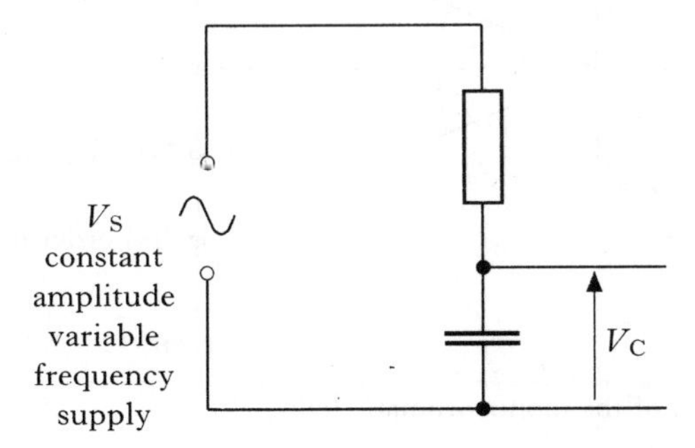

When the frequency of the supply is 100 Hz, the ratio V_C/V_S equals 0·5. The frequency of the supply voltage V_S is now increased to 1000 Hz while its amplitude is kept constant.

State whether the ratio V_C/V_S will increase, decrease or be unchanged. Justify your answer. **2**

Marks

36. A student uses the following method to determine the specific heat capacity of lead. He places some hot lead into a filter funnel containing ice at 0 °C as shown.

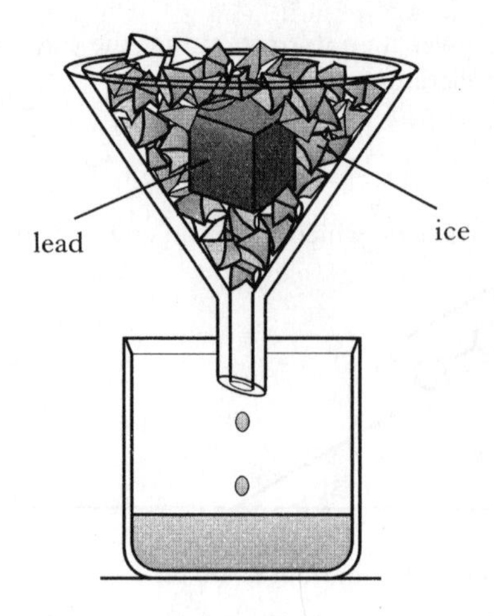

He measures the mass of water which collects in the beaker.

(*a*) What additional data would the student require to obtain a value for the specific heat capacity of lead?

(*b*) Using this method, the student obtains a value for the specific heat capacity of lead which is greater than the accepted value.

Suggest the reason for this difference. **3**

37. Monochromatic light is incident normally upon a grating which has 300 lines per mm. The angle between the two second order maxima is 46° as shown below.

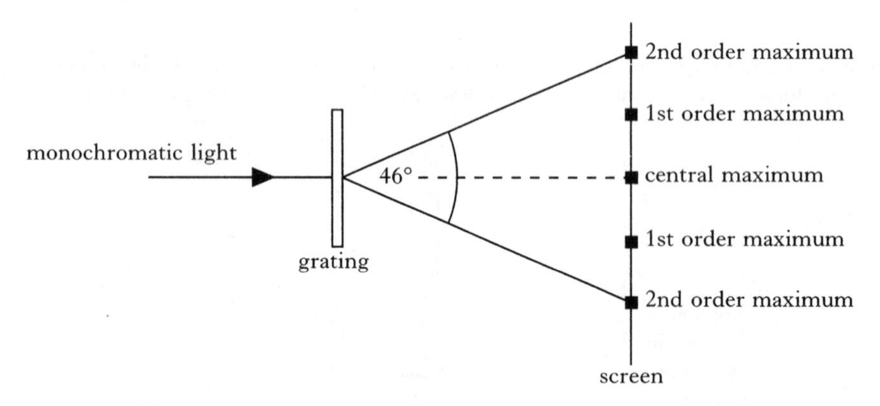

(*a*) Calculate the wavelength of the monochromatic light.

(*b*) What is the colour of this light? **3**

38. The following results were obtained for the half-life of a particular radioactive isotope:

53·0 s, 54·1 s, 57·5 s, 56·3 s, 55·1 s.

Calculate the best estimate of the half-life and the approximate random error in this value. **3**

[*END OF QUESTION PAPER*]

HIGHER PHYSICS ANSWERS

Paper I

Section A

E **2.** C **3.** B **4.** D **5.** A **6.** B **7.** B **8.** D **9.** E **10.** A
E **12.** C **13.** D **14.** D **15.** B **16.** C **17.** D **18.** D **19.** E **20.** C
D **22.** D **23.** A **24.** E **25.** D **26.** E **27.** E **28.** A **29.** A **30.** C

Section B

2·61 m s^{-2} **32.** 5·8 N **33.** 3·78 m^3 **34.** 60 J

Decreases $\left(V_C\ \alpha\ \frac{1}{f}\right)$

(a) Specific latent heat of ice; mass of lead; change in temperature of lead.
(b) Heat from surroundings also melted some ice, making mass of water in beaker too high.
(a) 651 nm *(b)* Red/Orange **38.** 55·2 ±0·9 s

SCOTTISH CERTIFICATE OF EDUCATION 1995

WEDNESDAY, 17 MAY
1.30 PM – 4.00 PM

PHYSICS
HIGHER GRADE
Paper II

Read carefully

1 All questions should be attempted.

2 Enter the question number clearly in the margin beside each question.

3 Any necessary data will be found in the Data Sheet.

4 Care should be taken not to give an unreasonable number of significant figures in the final answers to calculations.

5 Square-ruled paper (if used) should be placed inside the front cover of the answer book for return to the Scottish Qualifications Authority.

Marks

1. (*a*) State the difference between vector and scalar quantities. **1**

(*b*) In an orienteering event, competitors navigate from the start to control points around a set course.

Two orienteers, Andy and Paul, take part in a race in a flat area. Andy can run faster than Paul, but Paul is a better navigator.

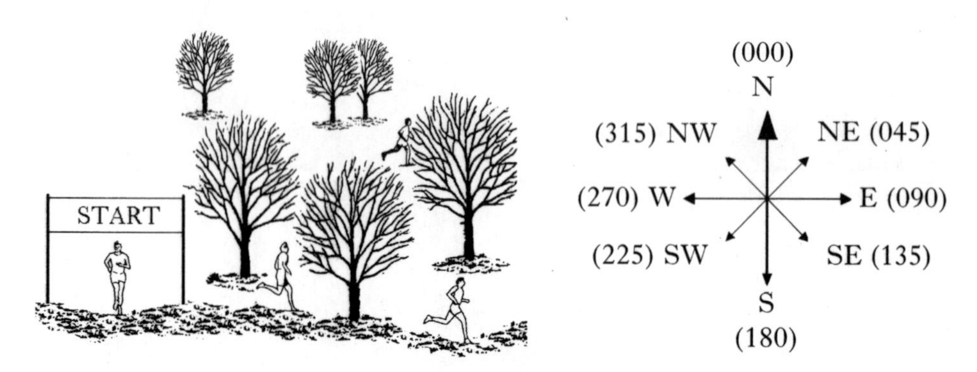

From the start, Andy runs 700 m north (000) then 700 m south-east (135) to arrive at the first control point. He has an average running speed of $3\ \text{m s}^{-1}$.

(i) By scale drawing or otherwise, find the displacement of Andy, from the starting point, when he reaches the first control point.

(ii) Calculate the average velocity of Andy between the start and the first control point.

(iii) Paul runs directly from the start to the first control point with an average running speed of $2{\cdot}5\ \text{m s}^{-1}$.

Determine the average velocity of Paul.

(iv) Paul leaves the starting point 5 minutes after Andy.

Show by calculation who is first to arrive at this control point. **9**

(10)

Marks

2. In a "handicap" sprint race, sprinters P and Q both start the race at the same time but from different starting lines on the track.

The handicapping is such that both sprinters reach the line XY, as shown below, at the same time.

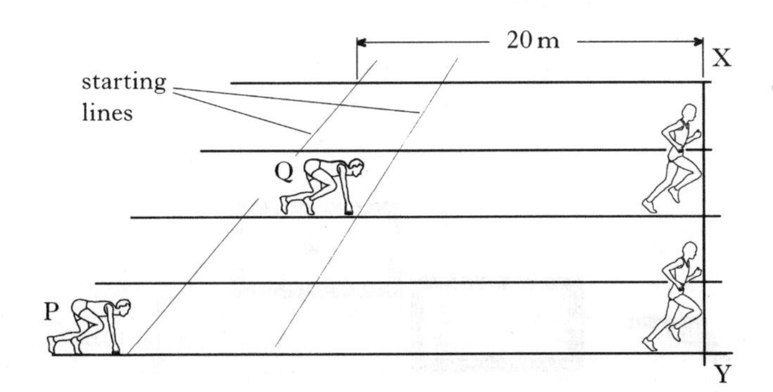

Sprinter P has a constant acceleration of $1{\cdot}6\,\text{m}\,\text{s}^{-2}$ from the start line to the line XY. Sprinter Q has a constant acceleration of $1{\cdot}2\,\text{m}\,\text{s}^{-2}$ from the start line to XY.

(*a*) Calculate the time taken by the sprinters to reach line XY. **2**

(*b*) Find the speed of **each** sprinter at this line. **3**

(*c*) What is the distance, in metres, between the starting lines for sprinters P and Q? **2**

(7)

Marks

3. (*a*) A bullet of mass 25 g is fired horizontally into a sand-filled box which is suspended by long strings from the ceiling. The combined mass of the bullet, box and sand is 10 kg.

After impact, the box swings upwards to reach a maximum height as shown in the diagram.

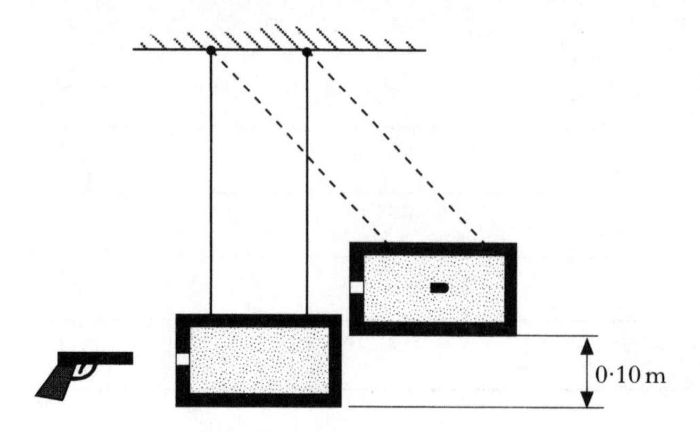

Calculate:

(i) the maximum velocity of the box after impact;

(ii) the velocity of the bullet just before impact. **5**

(*b*) The experiment is repeated with a metal plate fixed to one end of the box as shown.

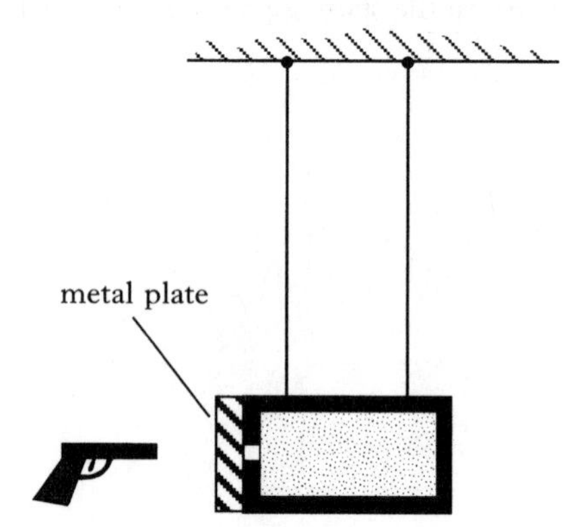

The mass of sand is reduced so that the combined mass of the sand, box and metal plate is 10 kg.

In this experiment, the bullet bounces back from the metal plate. Explain how this would affect the maximum height reached by the box compared with the maximum height reached in part (*a*). **2**

(7)

Marks

4. A crane is used to lower a concrete block of mass $5{\cdot}0 \times 10^3$ kg into the sea.

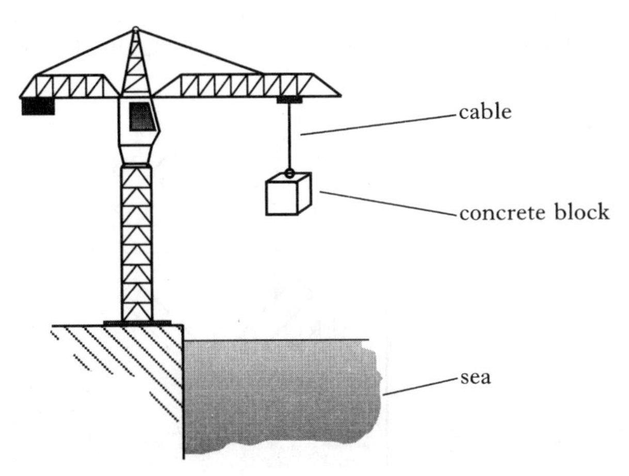

(*a*) The crane lowers the block towards the sea at a constant speed.
Calculate the tension in the cable supporting the block. **1**

(*b*) The crane lowers the block into the sea. The block is held stationary just below the surface of the sea as shown in the diagram below.

The tension in the cable is now $2{\cdot}9 \times 10^4$ N.

(i) Calculate the size of the buoyancy force acting on the block.

(ii) Explain how this buoyancy force is produced. **4**

(*c*) The block is now lowered to a greater depth.
What effect, if any, does this have on the tension in the cable?
Justify your answer. **2**

(7)

Marks

5. Four 10 Ω resistors R_1, R_2, R_3 and R_4 are connected in the form of a square ABCD. A fifth resistor R_5 of the same value is connected between A and C. This arrangement of resistors is connected in a circuit as shown below. The battery in the circuit has negligible internal resistance.

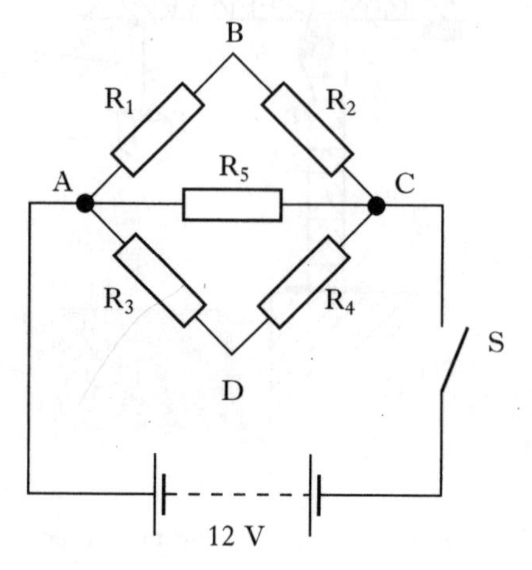

(*a*) Determine the total resistance between A and C. **2**

(*b*) The switch S is now closed.

(i) In which of the resistors is the greatest power developed?

(ii) Calculate the value of **this** power. **3**

(*c*) In a second experiment with the same resistors, the battery is connected across BD.

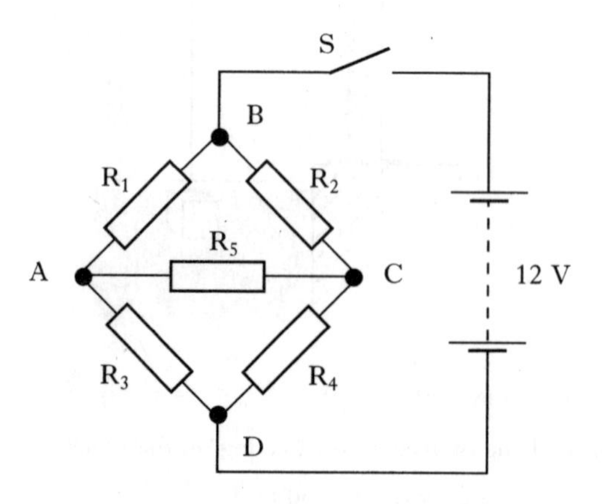

The switch S is now closed.

(i) Explain why there is no current in resistor R_5.

(ii) Calculate the current drawn from the battery. **4**

(9)

Marks

6. A certain car alarm system is triggered when the opening of a door of the car switches on the courtesy light.

The car alarm works by detecting the very small change in the voltage across the car battery which occurs when the courtesy light is switched on.

(*a*) The car battery has an e.m.f. of 12·0 V and an internal resistance r of 0·20 Ω as shown below.

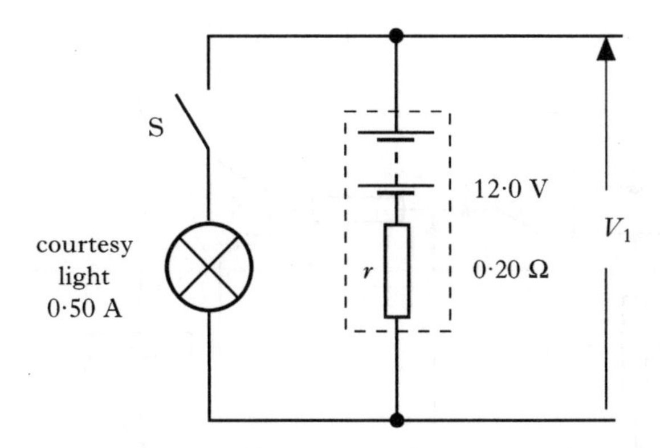

When the car door is opened, the switch S closes and the courtesy light draws a current of 0·50 A from the battery.

Show that the voltage V_1 across the battery falls to 11·9 V when the switch S is closed. **2**

(*b*) A capacitor and diode are **also** connected across the battery, as shown below, to provide a voltage V_2 of 12·0 V across the capacitor.

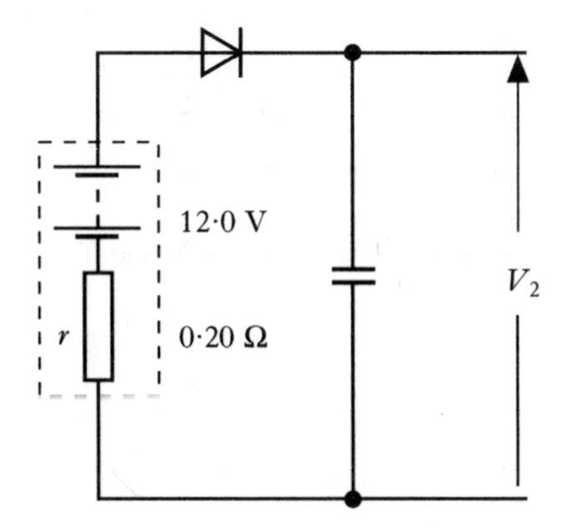

Explain why the voltage V_2 across the capacitor does not decrease from 12·0 V immediately after the car door is opened and the courtesy light comes on. **1**

Marks

(*c*) The voltages V_1 and V_2 obtained from the above circuits are then applied simultaneously as input voltages to the op-amp circuit shown below.

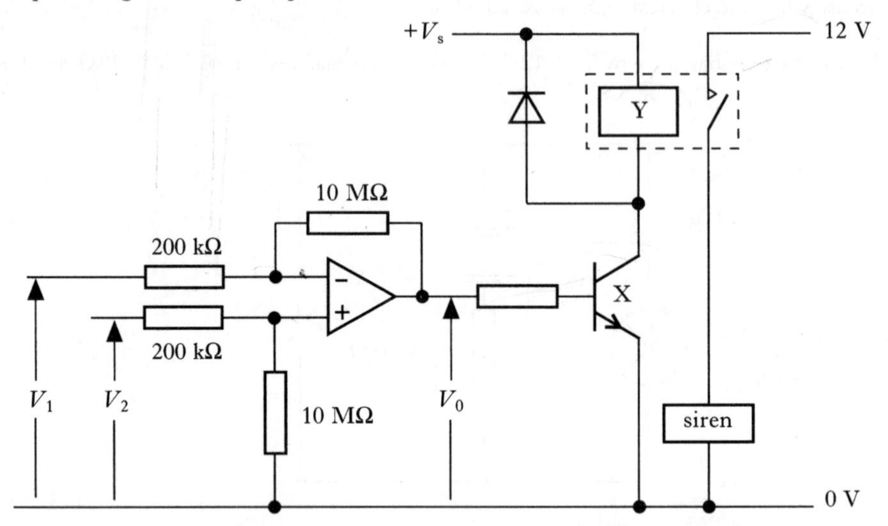

(i) In what mode is the op-amp being used in this circuit?

(ii) State the gain equation of the op-amp when operating in the mode shown above.

(iii) Hence calculate the output voltage V_0:

(A) while the car door is closed;

(B) when the car door is opened. **5**

(*d*) (i) Name the components labelled X and Y in the above circuit.

(ii) Describe how the output voltage V_0 of the op-amp can set off the siren of the car alarm.

Indicate in your answer the purpose of components X and Y. **4**

(12)

Marks

7. The circuit below is used to determine the internal resistance r of a battery of e.m.f. E.

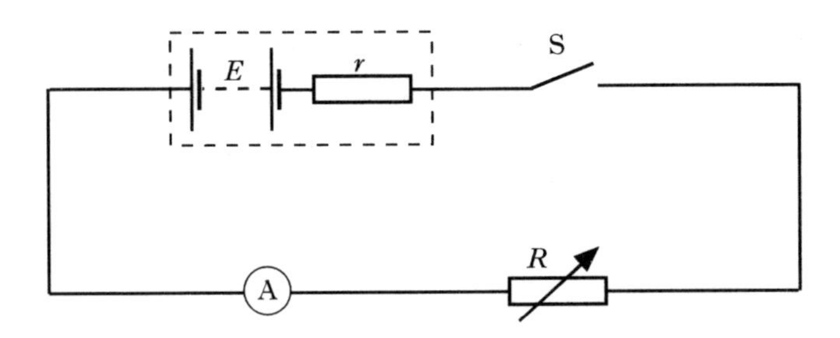

The variable resistor provides known values of resistance R.
For each value of resistance R, the switch S is closed and the current I is noted.

For each current, the value of $\frac{1}{I}$ is calculated.

In one such experiment, the following graph of R against $\frac{1}{I}$ is obtained.

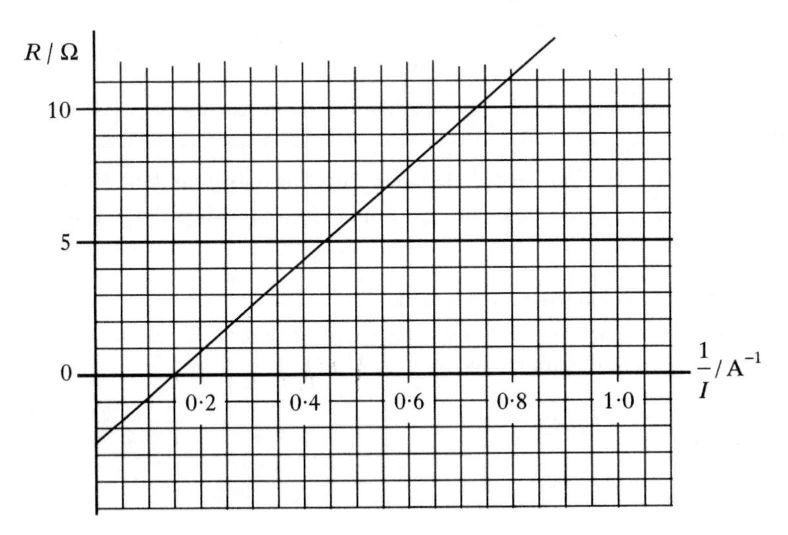

(*a*) Conservation of energy applied to the complete circuit gives the following relationship.

$$E = I(R + r)$$

Show that this relationship can be written in the form

$$R = \frac{E}{I} - r \; .$$ **1**

(*b*) Use information from the graph to find:

(i) the internal resistance of the battery;

(ii) the e.m.f. of the battery. **3**

(*c*) The battery is accidentally short-circuited.
Calculate the current in the battery when this happens. **2**

(6)

Marks

8. (*a*) It is quoted in a text book that
"the work function of caesium is $3{\cdot}04 \times 10^{-19}$ J".
Explain what is meant by the above statement. **1**

(*b*) In an experiment to investigate the photoelectric effect, a glass vacuum tube is arranged as shown below.

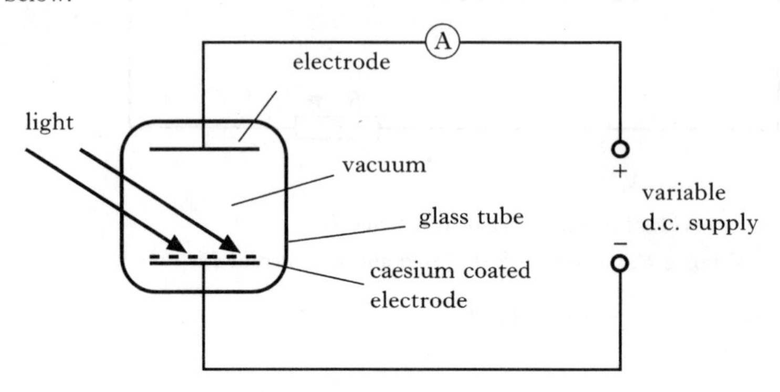

The tube has two electrodes, one of which is coated with caesium.
Light of frequency $6{\cdot}1 \times 10^{14}$ Hz is shone on to the caesium coated electrode.

(i) Calculate the maximum kinetic energy of a photoelectron leaving the caesium coated electrode.

(ii) An electron leaves the caesium coated electrode with this maximum kinetic energy.
Calculate its kinetic energy as it reaches the upper electrode when the p.d. across the electrodes is 0·8 V. **6**

(*c*) The polarity of the supply voltage is now reversed.
Calculate the minimum voltage which should be supplied across the electrodes to stop photoelectrons from reaching the upper electrode. **2**

(9)

Marks

9. A swimming pool is illuminated by a lamp built into the bottom of the pool.

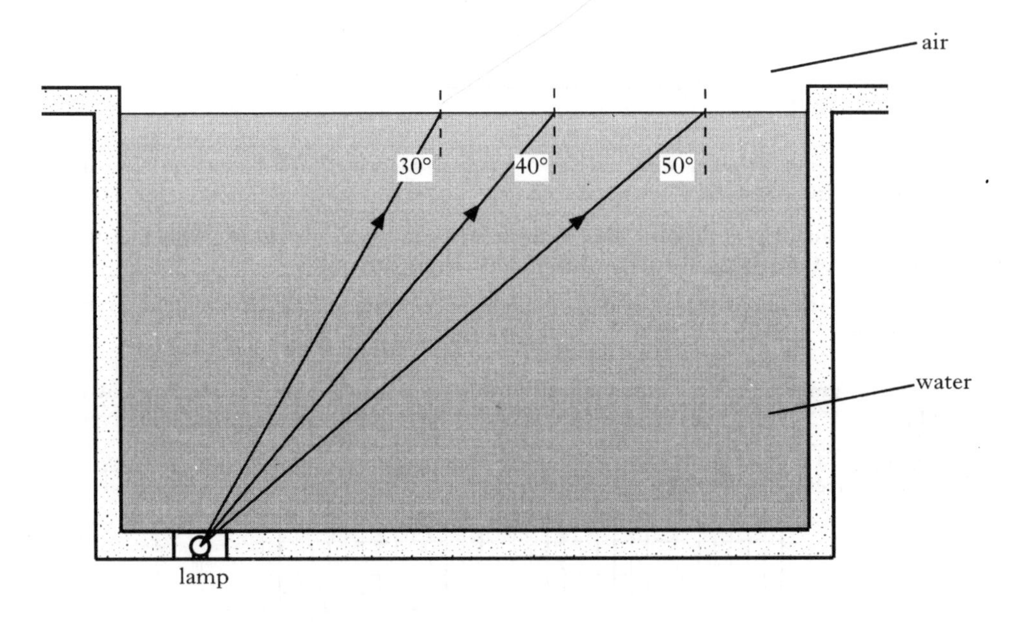

Three rays of light from the same point in the lamp are incident on the water-air boundary with angles of incidence of 30°, 40° and 50°, as shown above.

The refractive index of the water in the pool is 1·33.

(*a*) Draw a diagram to show clearly what happens to each ray at the boundary. Indicate on your diagram the sizes of appropriate angles.

All necessary calculations must be shown. **5**

(*b*) An observer stands at the side of the pool and looks into the water.

Explain, with the aid of a diagram, why the image of the lamp appears to be at a shallower depth than the bottom of the pool. **2**

(7)

Marks

10. (*a*) The diagram below represents the p–n junction of a light emitting diode (LED).

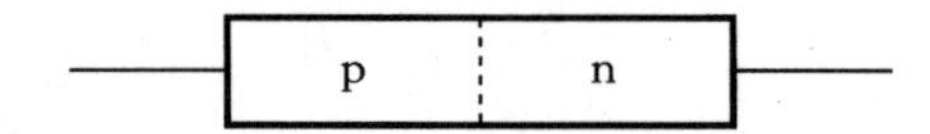

(i) Draw a diagram showing the above p–n junction connected to a battery so that the junction is forward biased.

(ii) When the junction is forward biased, there is a current in the diode. Describe the movement of the charge carriers which produces this current.

(iii) Describe how the charge carriers in the light emitting diode enable light to be produced. **5**

(*b*) The following graph shows the variation of current with voltage for a diode when it is forward biased.

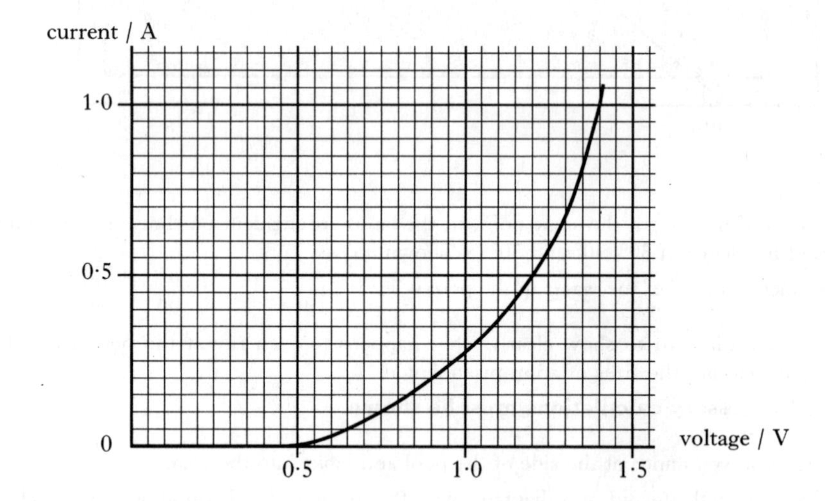

(i) What is the minimum voltage required for this diode to conduct?

(ii) What happens to the resistance of the diode as the voltage is increased above this minimum value?

Use information from the graph to justify your answer. **3**

(8)

Marks

11. The following statement represents a nuclear reaction which may form the basis of a nuclear power station of the future.

$$^{2}_{1}H + ^{3}_{1}H \rightarrow ^{4}_{2}He + ^{1}_{0}n$$

(*a*) State the name given to the above type of nuclear reaction. **1**

(*b*) Explain, using $E = mc^2$, how this nuclear reaction results in the production of energy. **2**

(*c*) Using the information given below, and any other data required from the Data Sheet, calculate the energy released in the above nuclear reaction.

mass of $^{3}_{1}H = 5 \cdot 00890 \times 10^{-27}$ kg

mass of $^{2}_{1}H = 3 \cdot 34441 \times 10^{-27}$ kg

mass of $^{4}_{2}He = 6 \cdot 64632 \times 10^{-27}$ kg

mass of $^{1}_{0}n = 1 \cdot 67490 \times 10^{-27}$ kg **3**

(*d*) Calculate how many of the reactions of the type represented above would occur each second to produce a power of 25 MW. **2**

(8)

[*END OF QUESTION PAPER*]

SOLUTIONS — PAPER II

Fully worked solutions to these questions are given in our book entitled *Solutions to Higher Grade Physics*

SCOTTISH CERTIFICATE OF EDUCATION 1996

FRIDAY, 17 MAY
9.30 AM – 11.00 AM

PHYSICS
HIGHER GRADE
Paper I

Read Carefully

1 All questions should be attempted.

2 The following data should be used when required unless otherwise stated.

Speed of light in vacuum c	$3{\cdot}00 \times 10^{8}$ m s^{-1}	Planck's constant h	$6{\cdot}63 \times 10^{-34}$ J s
Charge on electron e	$-1{\cdot}60 \times 10^{-19}$ C	Mass of electron m_e	$9{\cdot}11 \times 10^{-31}$ kg
Acceleration due to gravity g	$9{\cdot}8$ m s^{-2}	Mass of proton m_p	$1{\cdot}67 \times 10^{-27}$ kg

Section A (questions 1 to 30)

3 Check that the answer sheet is for Physics Higher I (Section A).

4 Answer the questions numbered 1 to 30 on the answer sheet provided.

5 Fill in the details required on the answer sheet.

6 Rough working, if required, should be done only on this question paper, or on the first two pages of the answer book provided—**not** on the answer sheet.

7 For each of the questions 1 to 30 there is only **one** correct answer and each is worth 1 mark.

8 Instructions as to how to record your answers to questions 1–30 are previously given.

Section B (questions 31 to 37)

9 Answer questions numbered 31 to 37 in the answer book provided.

10 Fill in the details on the front of the answer book.

11 Enter the question number clearly in the margin of the answer book beside each of your answers to questions 31 to 37.

12 Care should be taken **not** to give an unreasonable number of significant figures in the final answers to calculations.

1996

SECTION A

Answer questions 1–30 on the answer sheet.

1. A lift in a hotel makes a return journey from the ground floor to the top floor and then back again. The corresponding velocity-time graph is shown below.

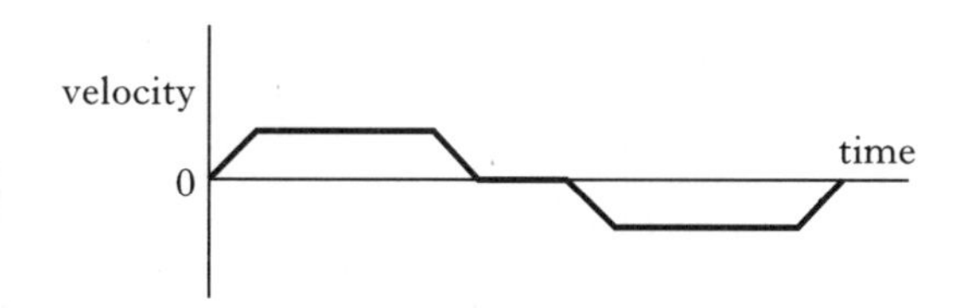

Which of the following shows the acceleration-time graph for the same journey?

A

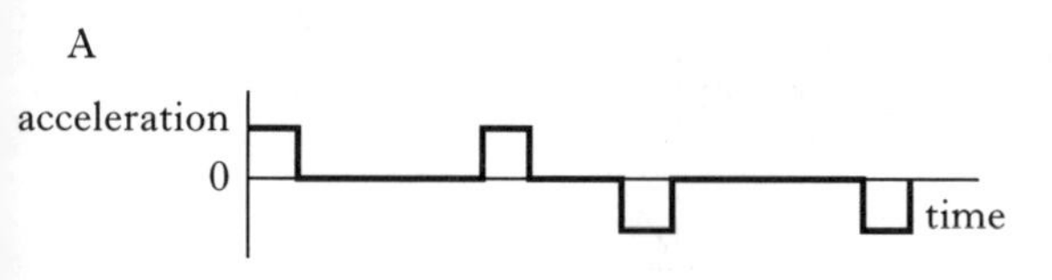

B

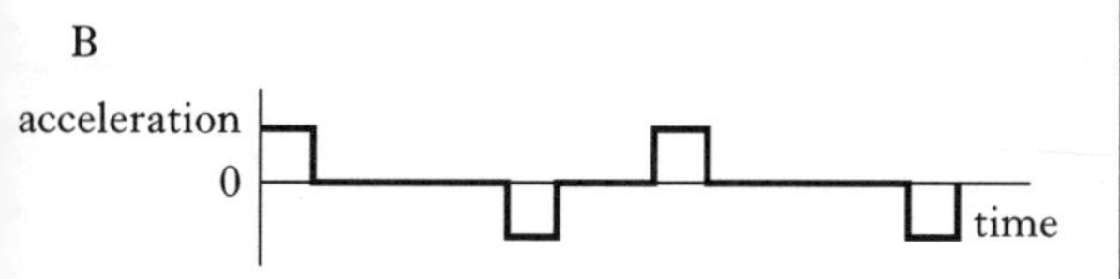

C

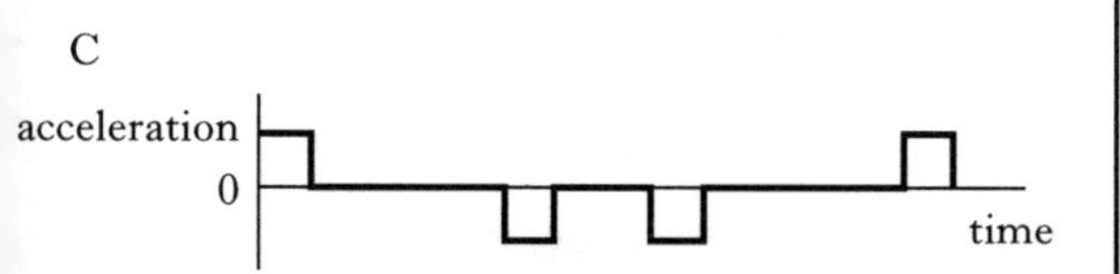

D

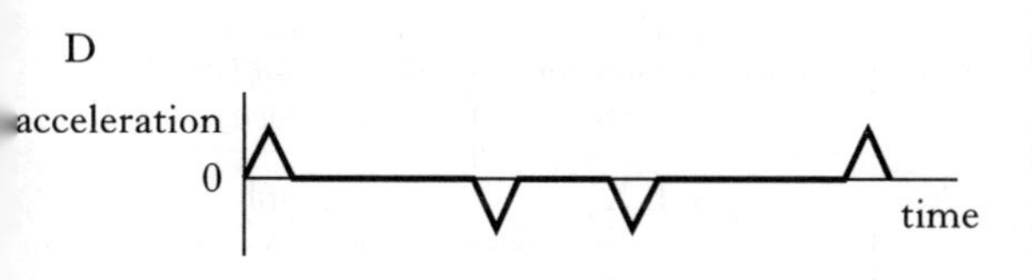

E

2. A car travels from X to Y and then it travels from Y to Z, as shown in the following diagram.

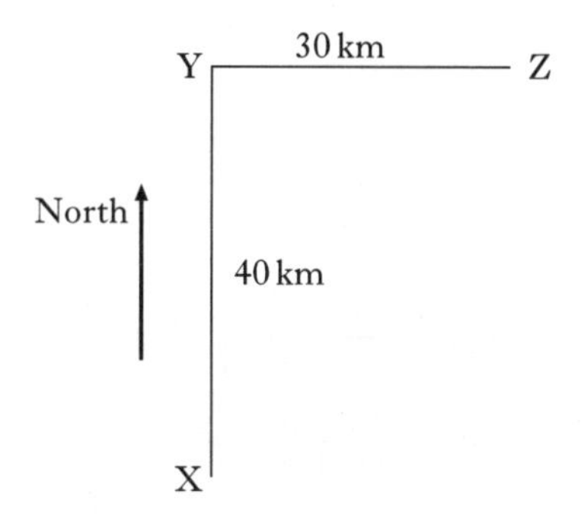

X to Y takes a time of one hour. Y to Z also takes one hour. Which of the following is a correct list of the magnitudes of the final displacement, average speed and average velocity for the complete journey?

	Displacement (km)	*Average speed* (km hr^{-1})	*Average velocity* (km hr^{-1})
A	50	35	35
B	70	35	25
C	50	35	25
D	70	70	50
E	50	70	25

3. An object attached to a parachute falls from a helicopter which is hovering at a height of 120 m above point X.

The object falls with a constant vertical component of velocity of value 12 m s^{-1}. A steady side-wind gives the object a constant horizontal component of velocity of value 5 m s^{-1}.

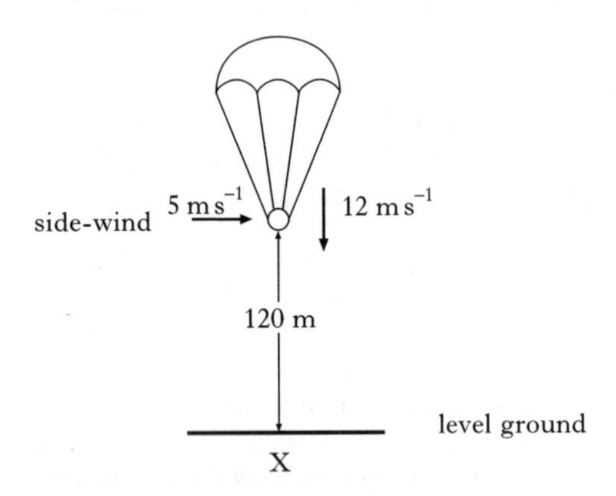

How far from point X does the object hit the ground?

A 24 m

B 50 m

C 60 m

D 120 m

E 150 m

4. A sledge is dragged at a **constant velocity** along the snow against a horizontal frictional force F. The rope pulling the sledge is at an angle of θ to the horizontal, as shown.

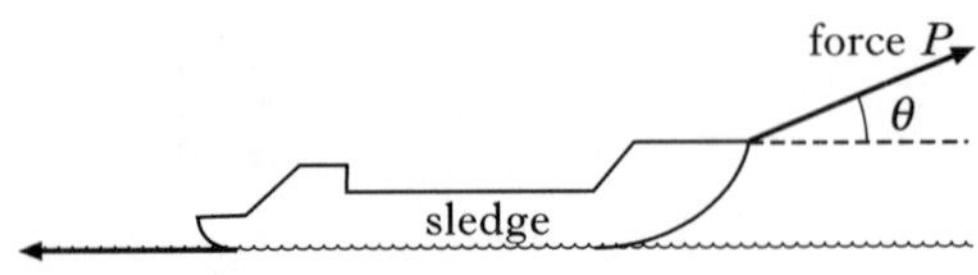

When the sledge is moving horizontally with a constant velocity, the force P pulling the rope is equal to

A F

B $F \cos \theta$

C $F \sin \theta$

D $\dfrac{F}{\cos \theta}$

E $\dfrac{F}{\sin \theta}$.

5. A ball is thrown horizontally over the edge of a cliff. When air resistance **is taken into account**, which graphs represent the horizontal and vertical components of the velocity of the ball during its flight?

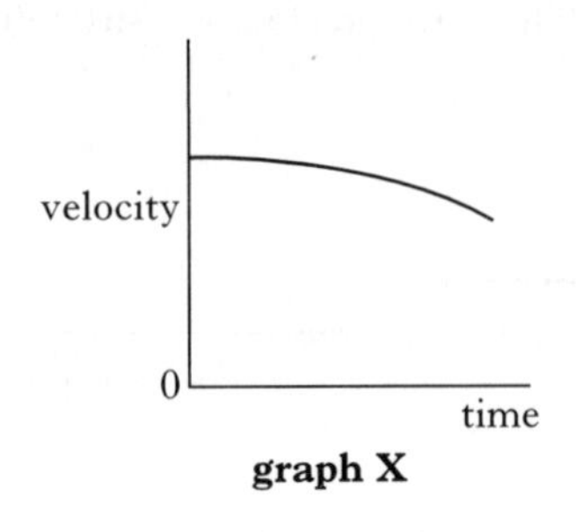

graph X

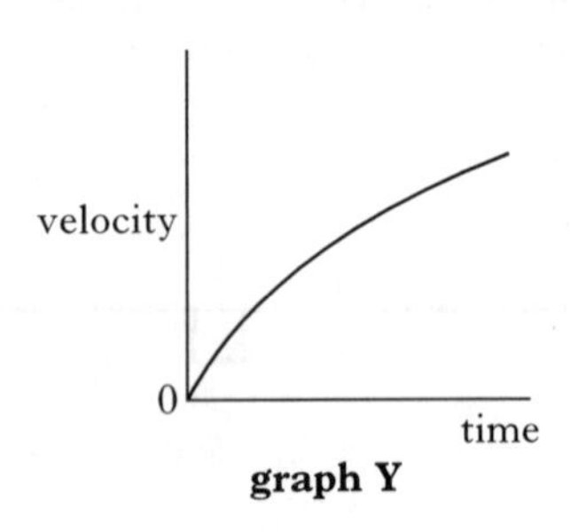

graph Y

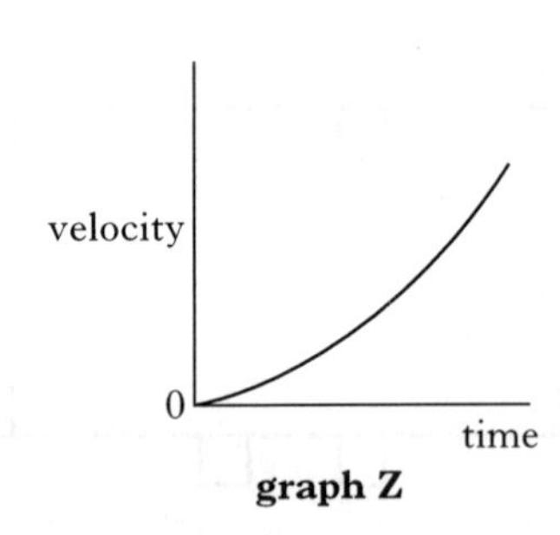

graph Z

	Horizontal component of velocity	*Vertical component of velocity*
A	graph X	graph X
B	graph X	graph Y
C	graph Y	graph X
D	graph Y	graph Z
E	graph Z	graph Z

6. A horizontal force of 20 N is applied as shown to two wooden blocks of masses 3 kg and 7 kg. The blocks are in contact with each other on a frictionless horizontal surface.

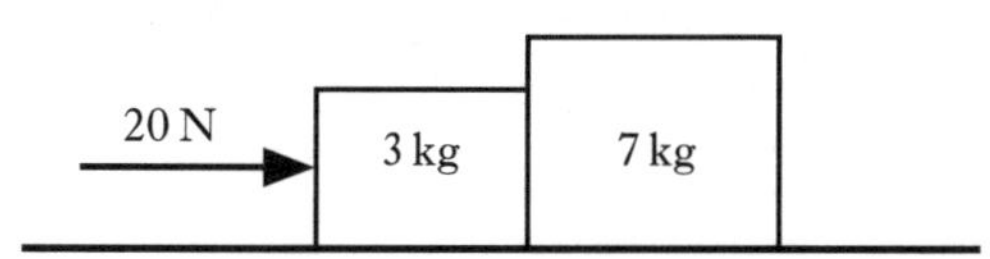

What is the size of the horizontal force acting on the 7 kg block?

A 20 N

B 14 N

C 10 N

D 8 N

E 6 N

7. An object of mass 1·0 kg hangs from a spring balance which is suspended on the inside of a small rocket, as shown below.

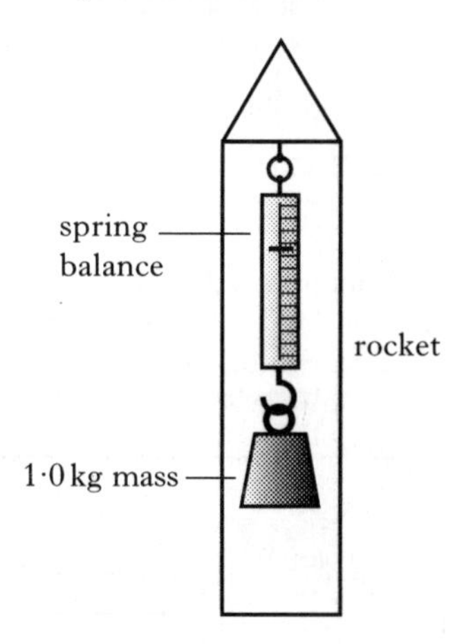

What is the reading on the balance when the rocket is accelerating upwards from the Earth's surface at 2·0 m s^{-2}? Use g = 9·8 m s^{-2}.

A 0 N

B 2·0 N

C 7·8 N

D 9·8 N

E 11·8 N

8. A field-gun of mass 1000 kg fires a shell of mass 10 kg with a velocity of 100 m s^{-1} East.

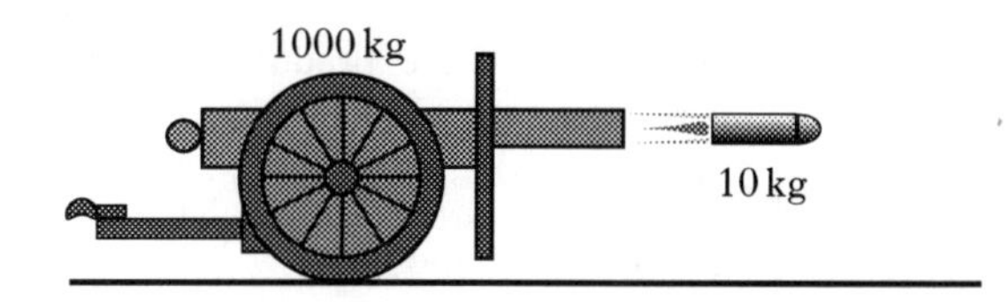

The velocity of the field-gun just after firing the shell is

A 0 m s^{-1}

B 1 m s^{-1} East

C 1 m s^{-1} West

D 10 m s^{-1} East

E 10 m s^{-1} West.

9. The graph below shows the force which acts on an object over a time interval of 8 seconds.

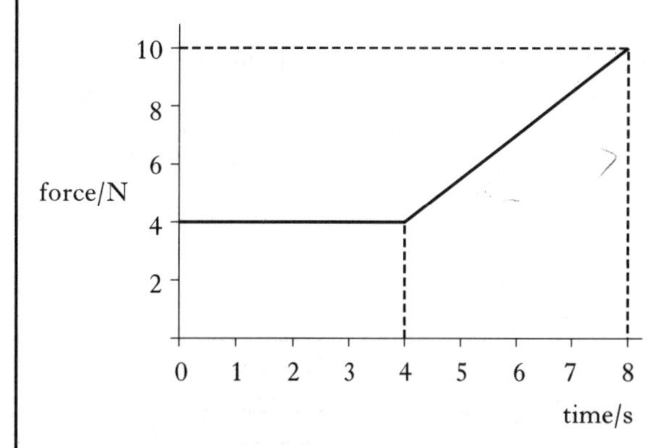

The momentum gained by the object during this 8 seconds is

A 12 N s

B 32 N s

C 44 N s

D 52 N s

E 72 N s.

10. An aircraft cruises at an altitude at which the air pressure is $0{\cdot}4 \times 10^5$ Pa. The inside of the aircraft cabin is maintained at a pressure of $1{\cdot}0 \times 10^5$ Pa. The area of an external cabin door is $2\ m^2$.

What is the outward force produced on this door by the pressures stated?

A $0{\cdot}3 \times 10^5$ N

B $0{\cdot}7 \times 10^5$ N

C $1{\cdot}2 \times 10^5$ N

D $2{\cdot}0 \times 10^5$ N

E $2{\cdot}8 \times 10^5$ N

11. The volt is equivalent to the

A farad/coulomb

B ampere/ohm

C joule/ampere

D joule/ohm

E joule/coulomb.

12. The diagram below illustrates a circuit in which the supply has an e.m.f. of 12 V and negligible internal resistance. Four load resistors, each of resistance 3 kΩ, are connected in the circuit as shown.

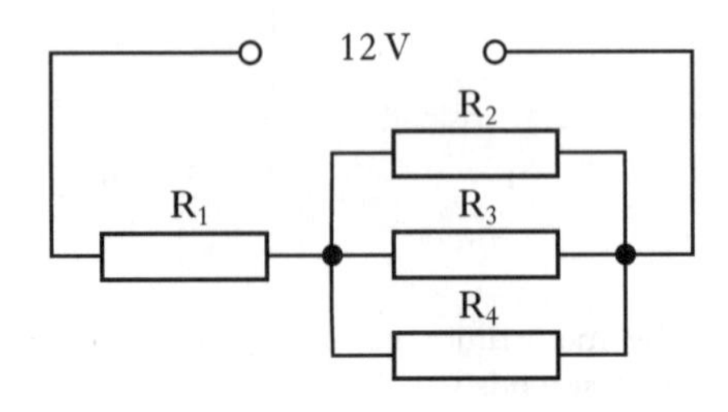

Which line in the table below indicates the potential differences in volts that would exist across the resistors?

	p.d. across R_1	*p.d. across* R_2	*p.d. across* R_3	*p.d. across* R_4
A	3 V	3 V	3 V	3 V
B	6 V	6 V	6 V	6 V
C	3 V	9 V	9 V	9 V
D	9 V	3 V	3 V	3 V
E	9 V	1 V	1 V	1 V

13. The circuit below can be used to determine the e.m.f. and the internal resistance of a cell.

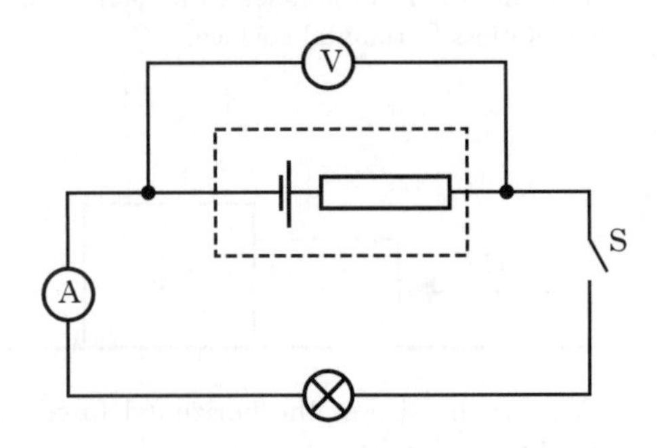

Ammeter and voltmeter readings are taken when switch S is open and again when it is closed. The results are as follows:

Switch S open: Current = zero : Voltage = V_1

Switch S closed: Current = I : Voltage = V_2

The e.m.f. of the cell is equal to

A V_1

B V_2

C $V_1 - V_2$

D $\frac{V_1}{I}$

E $\frac{(V_2 - V_1)}{I}$.

14. A battery has an e.m.f. of 6·0 V and an internal resistance of 2·0 Ω. It is connected to a 10·0 Ω resistor, as shown below.

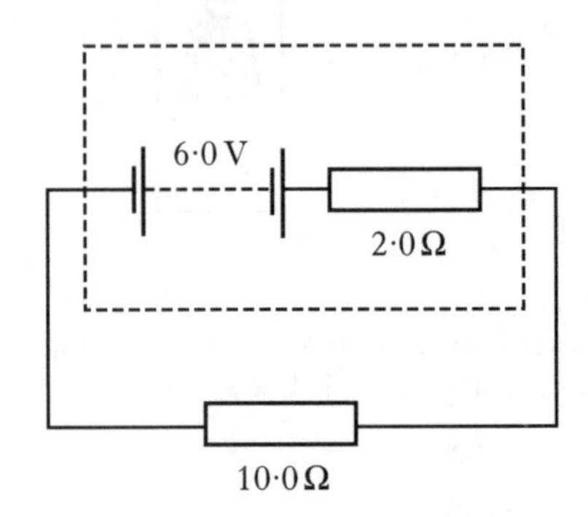

The p.d. across the 10·0 Ω resistor is

A 1·0 V

B 1·2 V

C 4·8 V

D 5·0 V

E 6·0 V.

15. The step-up transformer shown below is used to light a mains lamp at its correct rating. The input voltage to the primary is 6 V r.m.s. and the voltage across the lamp is 240 V r.m.s.

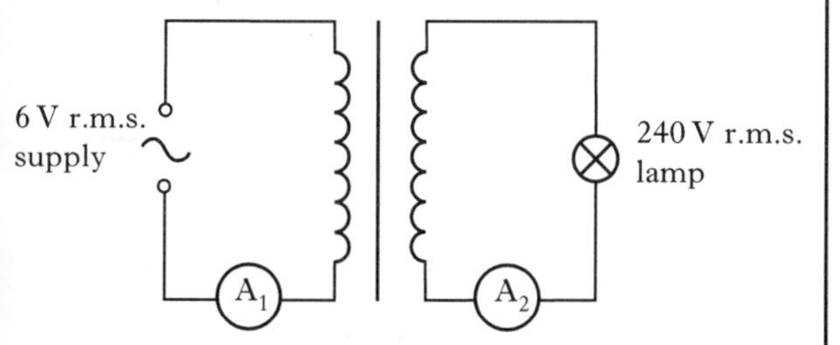

The a.c. ammeters A_1 and A_2 have negligible resistance.

The reading on A_1 is 5·0 A r.m.s. and the reading on A_2 is 0·1 A r.m.s.

The efficiency of the transformer is

A 2·5 %

B 40%

C 50%

D 80%

E 100%.

16. The circuit below shows two 6 Ω resistors connected in parallel to a 12 V d.c. supply of negligible resistance.

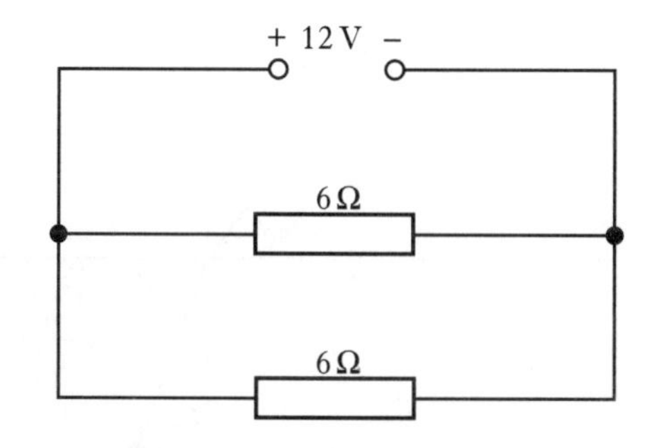

The total power developed in this circuit is

A 12 W

B 24 W

C 48 W

D 300 W

E 1200 W.

17. An 8 μF capacitor requires

A 8 μC to charge it to 1 V

B 1 μC to charge it to 8 V

C 8 μC to charge it to 8 V

D 8 C to charge it to 8 μV

E 1 C to charge it to 8 μV.

18. The following graph shows how the charge Q on a capacitor is related to the p.d. V applied across its plates.

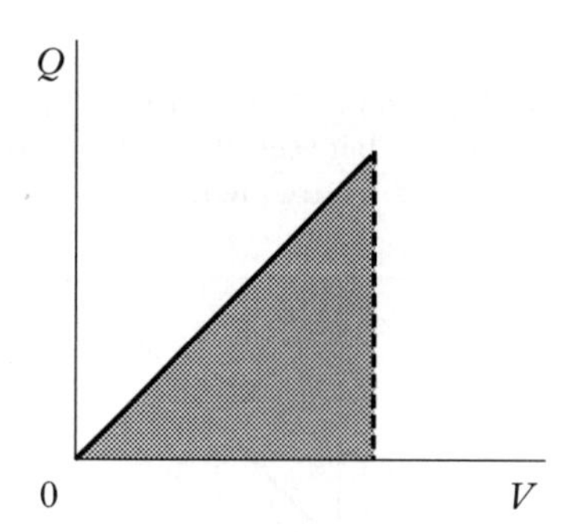

What does the shaded area under this graph represent?

A The distance between the plates of the capacitor

B The capacitance of the capacitor

C The working voltage of the capacitor

D The charge stored on the plates of the capacitor

E The energy stored in the capacitor

19. A resistor is connected in a circuit as shown. The output of the alternating supply can be varied in frequency but has a constant peak voltage.

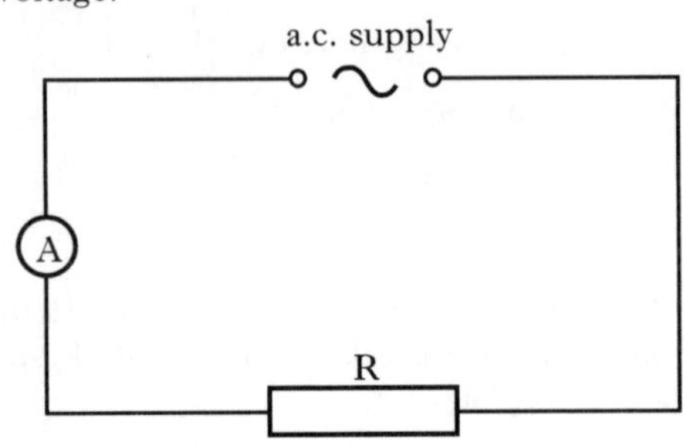

Which graph correctly represents the relationship between the r.m.s. current I in the resistor and the frequency f of the supply?

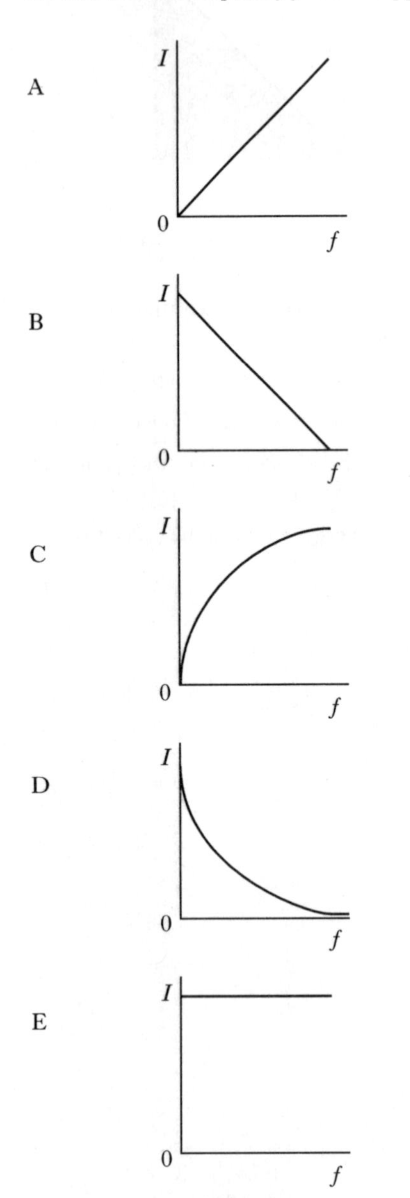

20. A capacitor is connected to a circuit as shown. The output of the alternating supply can be varied in frequency but has a constant peak voltage.

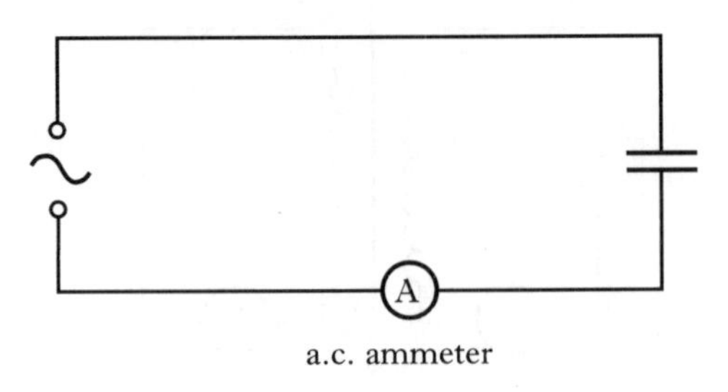

When the frequency of the output from the supply is steadily increased from 50 Hz to 5000 Hz, the reading on the a.c. ammeter will

A remain constant

B decrease steadily

C increase steadily

D increase then decrease

E decrease then increase.

21. The diagram shows a ray of light going into air from a crystalline substance.

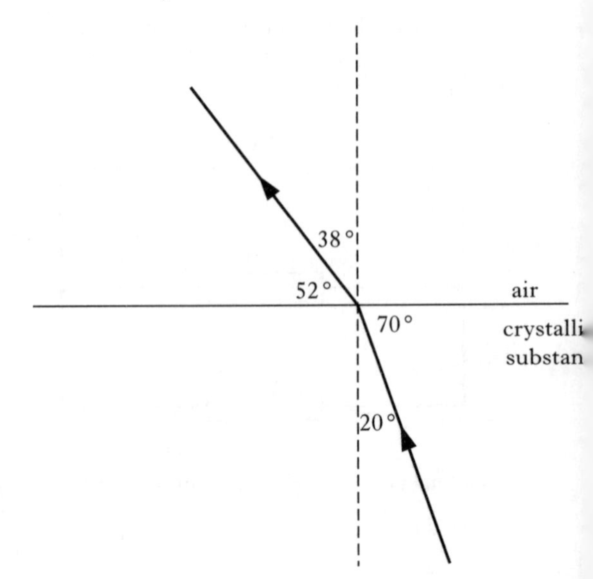

What is the refractive index of the crystalline substance?

A 1·2

B 1·3

C 1·8

D 1·9

E 2·3

22. A ray of monochromatic light passing from medium (1) into medium (2) follows the path PQR.

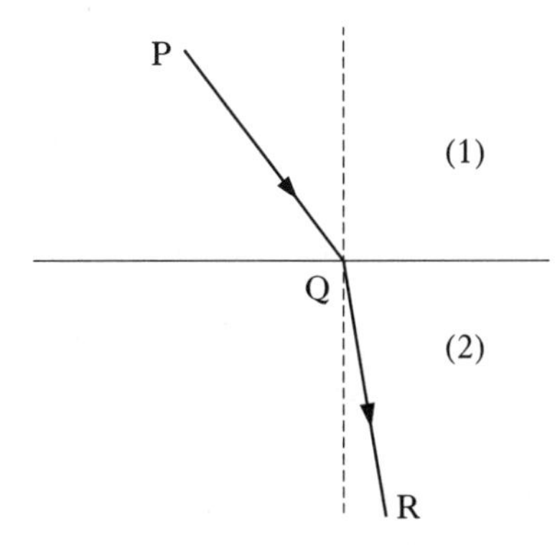

When the light passes from medium (1) to medium (2), its

A frequency is increased

B frequency is decreased

C wavelength is unchanged

D speed is increased

E speed is decreased.

23. Which one of the following diagrams shows the correct path for a ray of light travelling from air into a glass prism whose angles are 45 °, 90 ° and 45 °? The refractive index of the glass is 1·5.

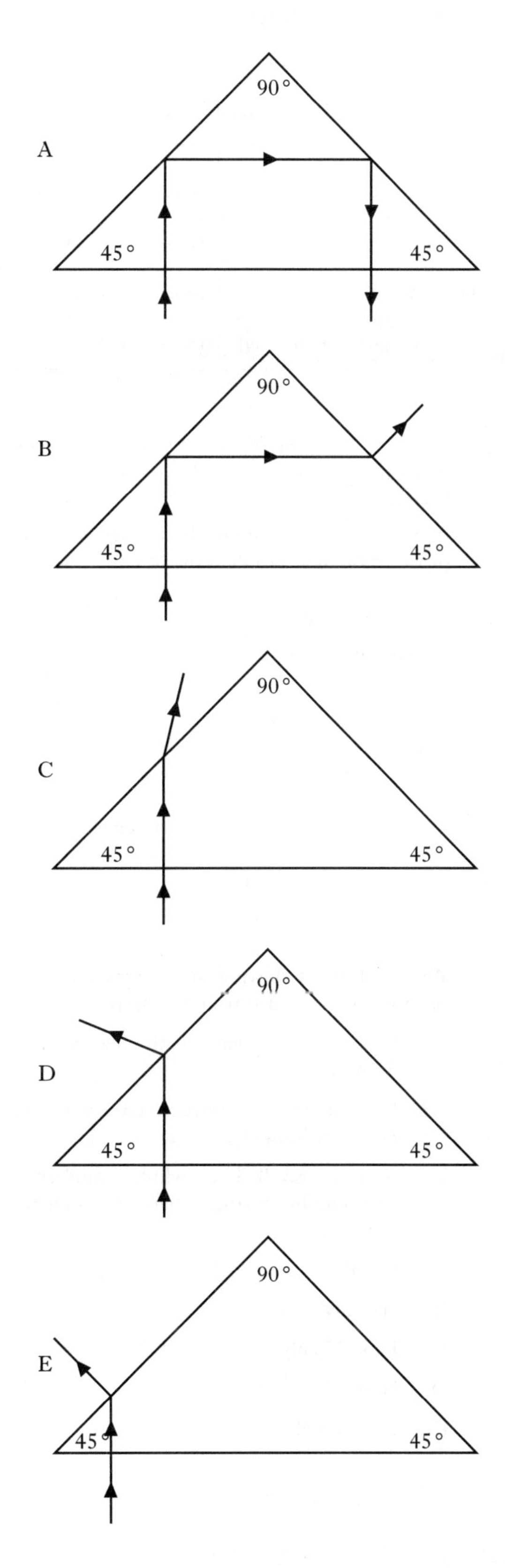

24. An n-type semiconductor is produced by adding arsenic impurity atoms to silicon. Which row in the following table describes the effect that this process has on the resistance and overall net charge of the material?

	Resistance	*Net Charge*
A	remains unchanged	remains unchanged
B	decreases	remains unchanged
C	increases	remains unchanged
D	decreases	more negative
E	remains unchanged	more negative

25. Microwaves of wavelength 2·8 cm pass through two narrow gaps G_1 and G_2 in an aluminium barrier.

Point P on the far side of the barrier is 11·2 cm further from one gap than the other.

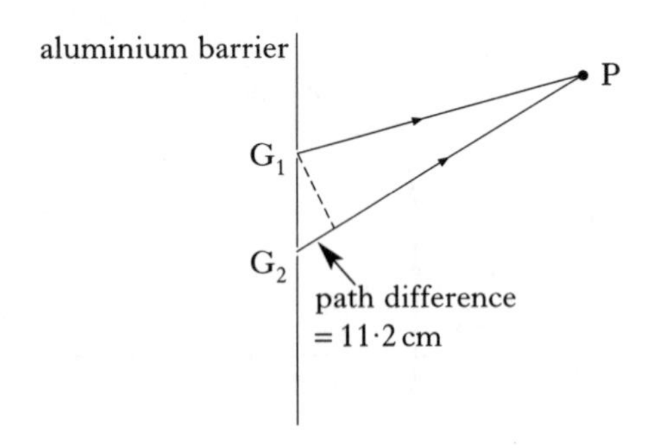

Which of the following statements about the radiation arriving at P from G_2 is/are true?

I It arrives in phase with the radiation from G_1.

II It combines constructively with the radiation from G_1.

III It has travelled a whole number of wavelengths further than the radiation from G_1.

A II only

B III only

C I and II only

D II and III only

E I, II and III

26. When a grating was set up to produce an interference pattern on a screen using a monochromatic light source, the fringes were too close together to allow accurate measurement.

Which **one** of the following changes would produce an increase in the separation of the fringes on the screen?

A Increasing the distance between the grating and the screen

B Using a grating with a greater separation between the lines on it

C Using another light source of shorter wavelength

D Using another light source of greater intensity

E Increasing the distance between the source and the grating

27. A small lamp is placed 1 metre above a desk. At a point on the desk directly below the lamp, the intensity of the light is I. The lamp may be treated as a point source of light.

The lamp is now raised until it is 2 metres above the desk. What is the new intensity of light at the same point on the desk?

A $\frac{I}{4}$

B $\frac{I}{2\sqrt{2}}$

C $\frac{I}{2}$

D $\frac{I}{\sqrt{2}}$

E $\sqrt{2}\, I$

28. The last two changes in a radioactive decay series are shown below.

A Bismuth nucleus emits a beta particle and its product, a Polonium nucleus, emits an alpha particle.

$$^{P}_{Q}\text{Bi} \xrightarrow[\text{decay}]{\beta} {}^{R}_{S}\text{Po} \xrightarrow[\text{decay}]{\alpha} {}^{208}_{82}\text{Pb}$$

Which numbers are represented by P, Q, R and S?

	P	Q	R	S
A	212	85	212	84
B	212	83	212	84
C	211	85	207	86
D	210	83	208	81
E	210	83	210	84

29. The diagram below shows the energy levels in an atom.

$-5{\cdot}2 \times 10^{-19}$ J ———— E_3

$-9{\cdot}0 \times 10^{-19}$ J ———— E_2

$-16{\cdot}4 \times 10^{-19}$ J ———— E_1

$-24{\cdot}6 \times 10^{-19}$ J ———— E_0

An electron is excited from energy level E_2 to level E_3 by absorbing energy. What is the frequency of light being used to excite the electron?

A $1{\cdot}74 \times 10^{-15}$ Hz

B $5{\cdot}73 \times 10^{14}$ Hz

C $1{\cdot}69 \times 10^{15}$ Hz

D $2{\cdot}14 \times 10^{15}$ Hz

E $2{\cdot}92 \times 10^{15}$ Hz

30. A detector placed near a source of gamma rays records a count rate of 480 counts per second.

A slab of material of thickness 3 cm is then placed between the source and the detector. The half-value thickness of this material is 1 cm and the half-life of the source is 1 day.

After 1 day, what is the count rate recorded by the detector ?

A 240 counts per second

B 160 counts per second

C 80 counts per second

D 60 counts per second

E 30 counts per second

SECTION B

Write your answers to questions 31 to 37 in the answer book.

Marks

31. An archer fires an arrow at a target which is 30 m away.

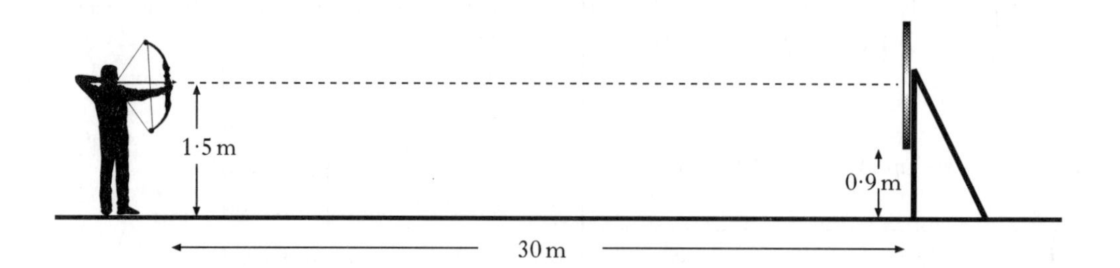

The arrow is fired horizontally from a height of 1·5 m and leaves the bow with a velocity of 100 $m\,s^{-1}$.

The bottom of the target is 0·9 m above the ground.

Show by calculation that the arrow hits the target. Use $g = 9{\cdot}8\ m\,s^{-2}$. **3**

32. A mooring buoy is tethered to the seabed by a rope which is too short. The buoy floats under the water at high tide. The weight of the buoy is 50 N.

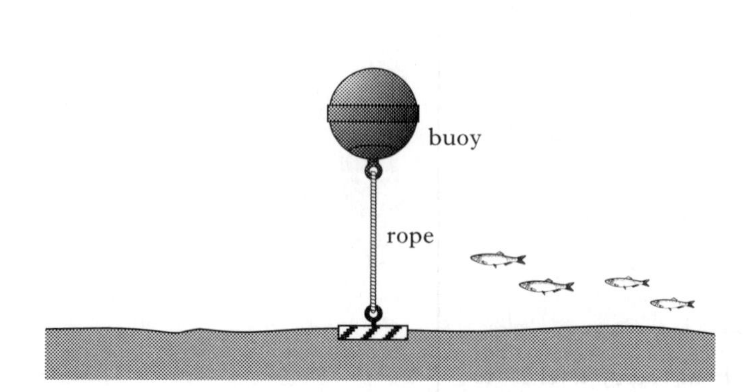

(*a*) (i) Draw a labelled diagram to show all the forces acting on the buoy in the vertical direction.

(ii) The tension in the rope is 1200 N.

Calculate the buoyancy force.

(*b*) The rope now snaps and the buoy starts to rise.

What is the size of the buoyancy force on the buoy when it is just below the surface of the water? **4**

Marks

33. A cell of e.m.f. 1·5 V and internal resistance 0·2 Ω is connected across a lamp.
A second identical lamp is now connected in parallel with the first lamp.
Describe and explain what happens to the brightness of the first lamp. **2**

34. The peak value of an a.c. voltage is 12 V.
Calculate
(*a*) the r.m.s. voltage
(*b*) the power dissipated in a 4 Ω resistor by the r.m.s. voltage. **3**

35. Light of wavelength 600 nm is passed through a grating.
The grating has $2{\cdot}5 \times 10^5$ lines per metre.
Calculate the angle at which the first maximum appears. **2**

36. A particular atom has energy levels as shown below.

E_3

E_2

E_1

Ground state

Transitions are possible between all these levels to produce emission lines in the spectrum.
(*a*) How many lines are in the spectrum of this atom?
(*b*) Between which two energy levels would an electron transition lead to the emission of radiation of the lowest frequency?
(*c*) Explain why some lines in the spectrum are more intense than others. **3**

37. The work function for sodium metal is $2{\cdot}9 \times 10^{-19}$ J.
Light of wavelength $5{\cdot}4 \times 10^{-7}$ m strikes the surface of this metal. What is the kinetic energy of the electrons emitted from the surface? **3**

[*END OF QUESTION PAPER*]

1996

HIGHER PHYSICS ANSWERS
Paper I

Section A

1. C	**2.** C	**3.** B	**4.** D	**5.** B	**6.** B	**7.** E	**8.** C	**9.** C	**10.**
11. E	**12.** D	**13.** A	**14.** D	**15.** D	**16.** C	**17.** A	**18.** E	**19.** E	**20.**
21. C	**22.** E	**23.** A	**24.** B	**25.** E	**26.** A	**27.** A	**28.** B	**29.** B	**30.**

Section B

31. Flight time = 0·3 s
Distance arrow falls in this time $= \frac{1}{2} g t^2$
$= 0{\cdot}44$
Thus distance < 0·6 m so arrow hits target.

32. *(a)* (i) (ii) 1250 N (b) 1250 N

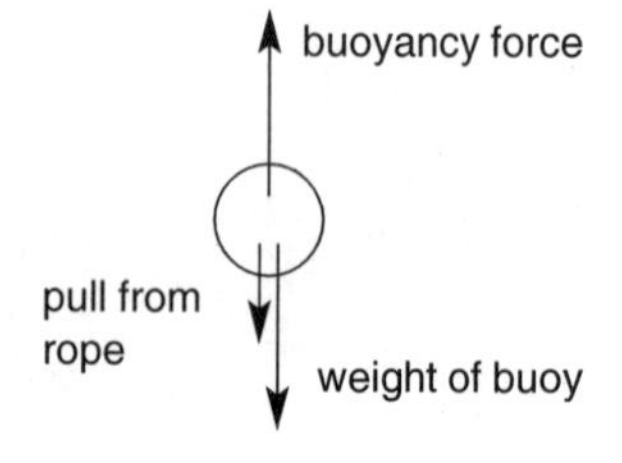

33. Brightness decreases. Larger current reduces the terminal p.d. Voltage across lamp is reduced. Power of lamp is reduc

34. *(a)* 8·48 V *(b)* 17·9 W

35. 8·6°

36. *(a)* 6 *(b)* E_3 to E_2 *(c)* More electrons make these transitions in a given time.

37. $7{\cdot}8 \times 10^{-20}$ J

SCOTTISH CERTIFICATE OF EDUCATION 1996

FRIDAY, 17 MAY
1.30 PM – 4.00 PM

PHYSICS HIGHER GRADE Paper II

Read carefully

1 All questions should be attempted.

2 Enter the question number clearly in the margin beside each question.

3 Any necessary data will be found in the Data Sheet.

4 Care should be taken not to give an unreasonable number of significant figures in the final answers to calculations.

5 Square-ruled paper (if used) should be placed inside the front cover of the answer book for return to the Scottish Qualifications Authority.

Marks

1. The manufacturers of tennis balls require that the balls meet a given standard.

When dropped from a certain height onto a test surface, the balls must rebound to within a limited range of heights.

The ideal ball is one which, when dropped from rest from a height of 3·15 m, rebounds to a height of 1·75 m as shown below.

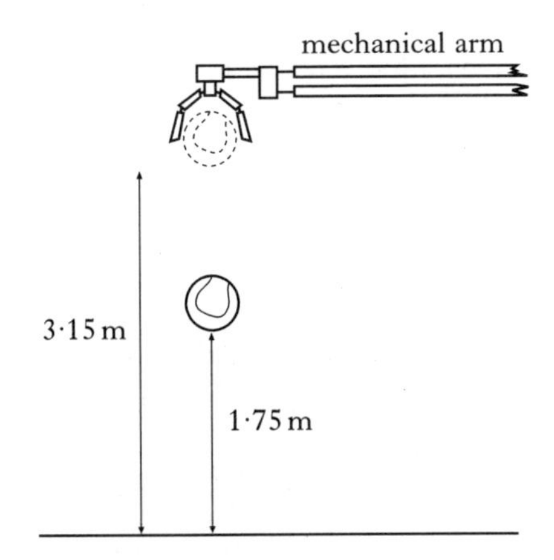

(*a*) Assuming air resistance is negligible, calculate

(i) the speed of an ideal ball just before contact with the ground

(ii) the speed of this ball just after contact with the ground. **3**

(*b*) When a ball is tested six times, the rebound heights are measured to be

1·71 m, 1·78 m, 1·72 m, 1·76 m, 1·73 m, 1·74 m.

Calculate

(i) the mean value of the height of the bounce

(ii) the random error in this value. **3**

(6)

Marks

2. A child on a sledge slides down a slope which is at an angle of 20° to the horizontal as shown below.

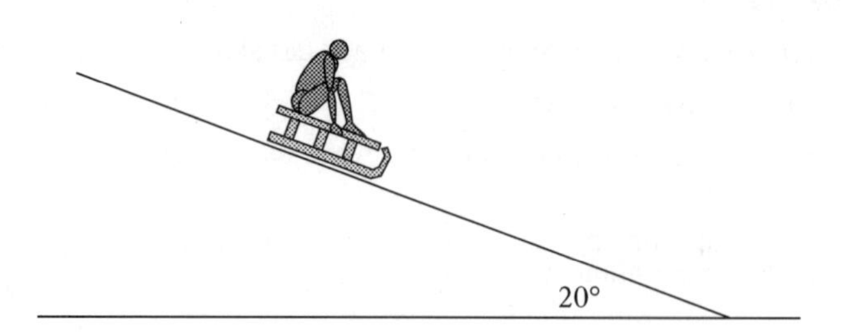

The combined weight of the child and the sledge is 400 N. The frictional force acting on the sledge and child at the start of the slide is 20·0 N.

(*a*) (i) Calculate the component of the combined weight of the child and sledge down the slope.

(ii) Calculate the initial acceleration of the sledge and child. **4**

(*b*) The child decides to start the slide from further up the slope. Explain whether or not this has any effect on the initial acceleration. **2**

(*c*) During the slide, the sledge does not continue to accelerate but reaches a constant speed. Explain why this happens. **2**

(8)

Marks

3. During a test on car safety, two cars as shown below are crashed together on a test track.

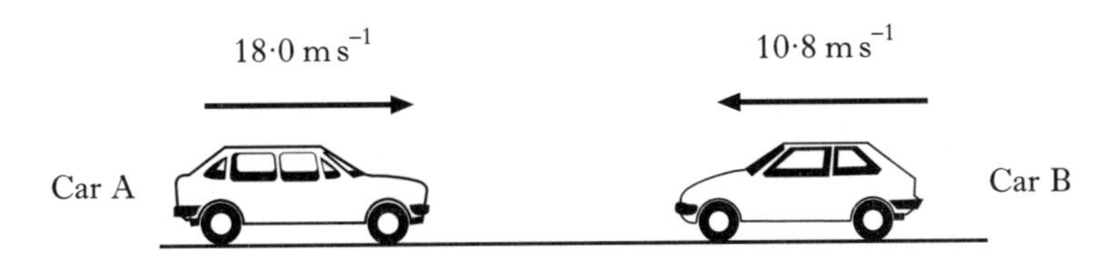

(*a*) Car A, which has a mass of 1200 kg and is moving at $18{\cdot}0\ \mathrm{m\,s^{-1}}$, approaches car B, which has a mass of 1000 kg and is moving at $10{\cdot}8\ \mathrm{m\,s^{-1}}$, in the opposite direction.

The cars collide head on, lock together and move off in the direction of car A.

(i) Calculate the speed of the cars immediately after the collision.

(ii) Show by calculation that this collision is inelastic. **4**

(*b*) During a second safety test, a dummy in a car is used to demonstrate the effects of a collision. During the collision, the head of the dummy strikes the dashboard at $20\ \mathrm{m\,s^{-1}}$ as shown below and comes to rest in 0·02 s.

The mass of the head is 5 kg.

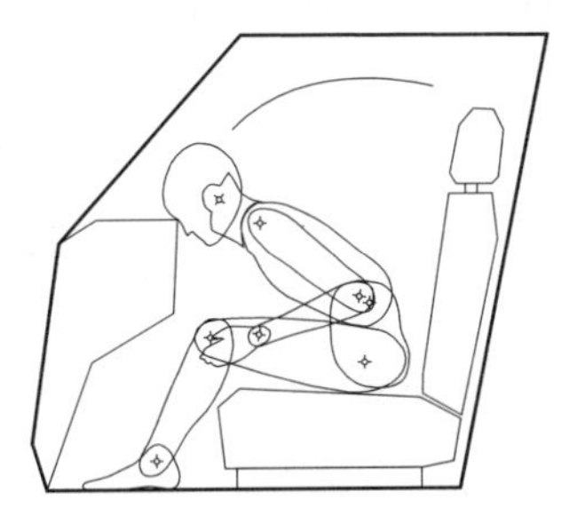

(i) Calculate the average force exerted by the dashboard on the head of the dummy during the collision.

(ii) If the contact area between the head and the dashboard is $5 \times 10^{-4}\ \mathrm{m^2}$, calculate the pressure which this force produces on the head of the dummy.

(iii) The test on the dummy is repeated with an airbag which inflates during the collision. During the collision, the head of the dummy again travels forward at $20\ \mathrm{m\,s^{-1}}$ and is brought to rest by the airbag.

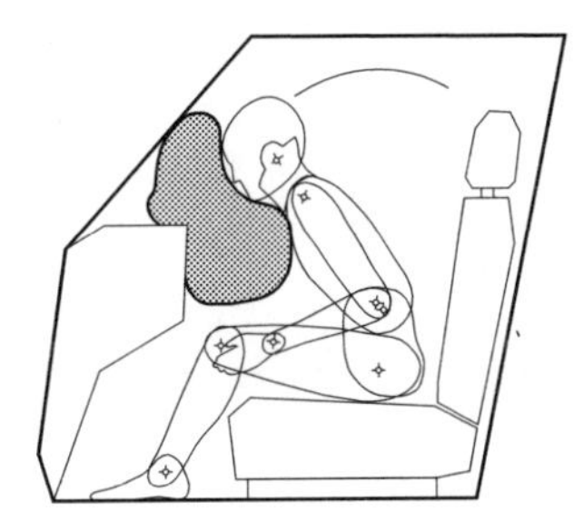

Explain why there is less risk of damage to the head of the dummy when the airbag is used. **5**

(9)

Marks

4. The diagram below illustrates an experiment to investigate the relationship between pressure and volume of a gas.

The apparatus consists of a calibrated syringe fitted with a gas-tight piston. Air is trapped in the syringe and the pressure of the trapped air can be monitored using a pressure sensor and a meter.

The pressure of the trapped air can be altered by exerting a force on the piston.

The temperature of the trapped air is assumed to be constant during the experiment.

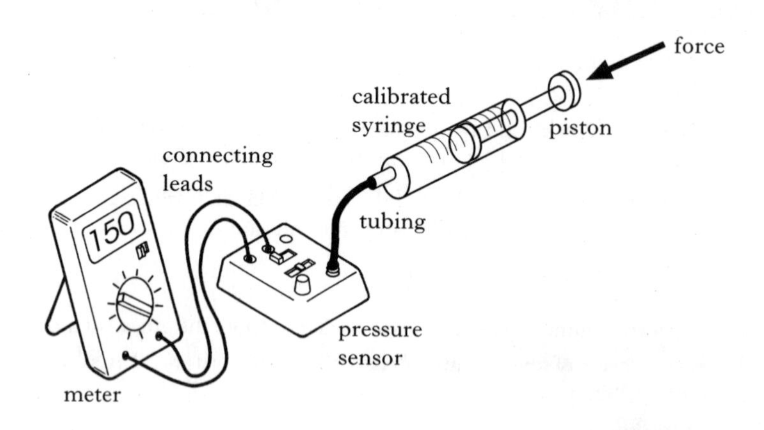

The following measurements of pressure and volume are recorded.

Pressure/kPa	100	150	200	250
Volume/cm^3	14·7	9·9	7·4	5·9

(*a*) Using all the data, establish the relationship between the pressure and volume of the trapped air. **2**

(*b*) The force on the piston is now altered until the volume of the trapped air is 5·0 cm^3.
Calculate the pressure of the trapped air. **2**

(*c*) The force is now removed from the piston.
Explain the subsequent motion of the piston in terms of the movement of the air molecules. **2**

(*d*) The tubing between the syringe and the pressure sensor is replaced by one of longer length. What effect would this have on the results of the experiment? **1**

(7)

Marks

5. A Wheatstone bridge is used to monitor the temperature of gas in a pipe.

A length of platinum resistance wire forms one part of the Wheatstone bridge circuit. The wire is inserted into the pipe containing the gas as shown below. The 9 V supply has negligible internal resistance.

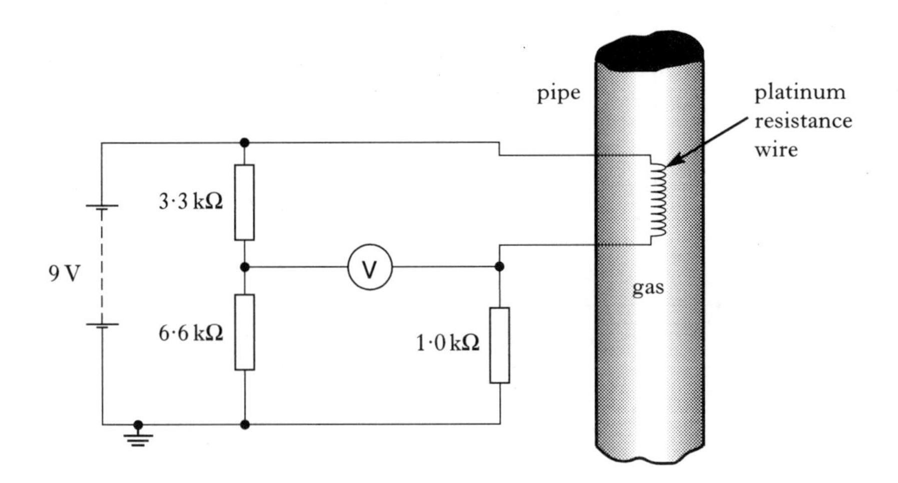

(*a*) (i) The bridge is initially balanced. What is the reading on the voltmeter?

(ii) Calculate the resistance of the platinum wire. **3**

(*b*) The graph below shows how the resistance of the platinum wire varies with temperature.

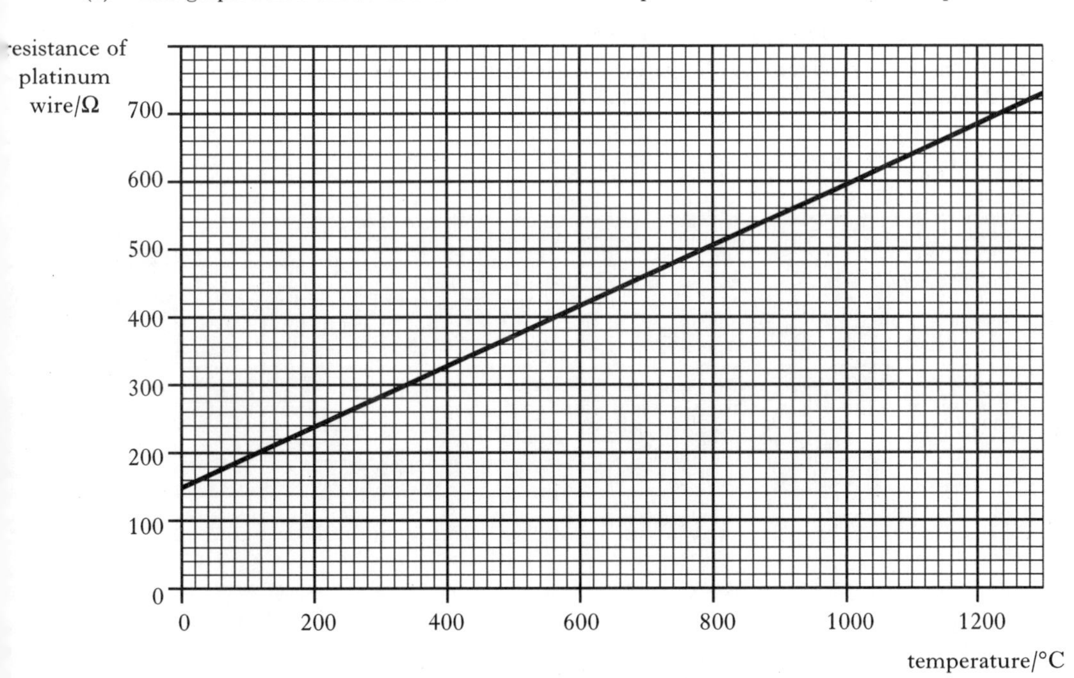

The temperature of the gas and the platinum wire is changed to 600 °C. The Wheatstone bridge is now out of balance.

(i) What is the resistance of the platinum wire at 600 °C?

(ii) Calculate the p.d. across the 1·0 kΩ resistor.

(iii) Calculate the reading on the voltmeter. **6**

(9)

Marks

6. A particle accelerator increases the speed of protons by accelerating them between a pair of parallel metal plates, **A** and **B**, connected to a power supply as shown below.

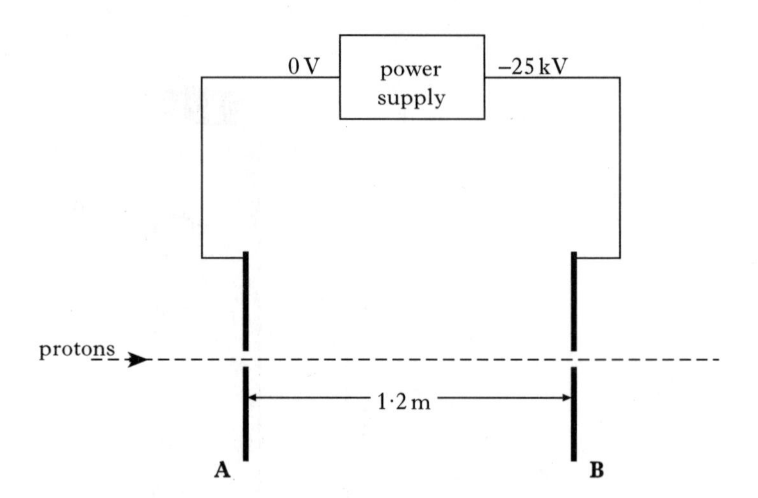

The potential difference between **A** and **B** is 25 kV.

(*a*) Show that the kinetic energy gained by a proton between plates **A** and **B** is $4 \cdot 0 \times 10^{-15}$ J. **2**

(*b*) The kinetic energy of a proton at plate **A** is $1 \cdot 3 \times 10^{-16}$ J.
Calculate the velocity of the proton on reaching plate **B**. **3**

(*c*) The plates are separated by a distance of 1·2 m.
Calculate the force produced by the particle accelerator on a proton as it travels between plates **A** and **B**. **2**

(7)

Marks

7. A student is investigating the charging and discharging of a 10 000 μF capacitor using the circuit shown below. The 6 V supply has negligible internal resistance.

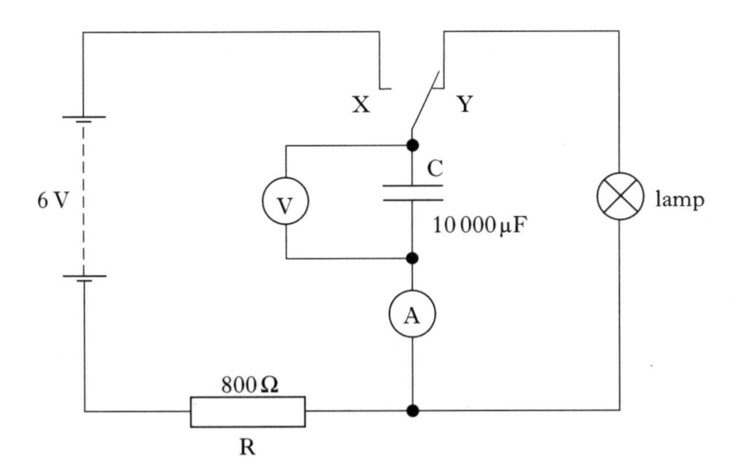

Initially the capacitor is uncharged and the switch is in position Y. The switch is moved to position X until the capacitor is fully charged and then finally back to position Y.

The graphs below show the p.d. V_C across the capacitor and the current I_C in the ammeter during this process.

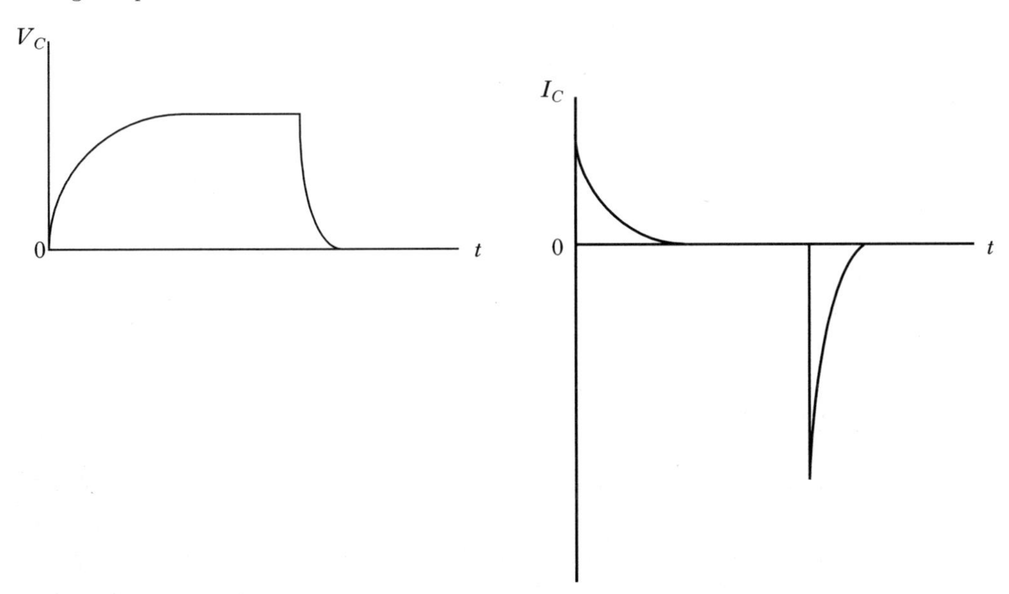

(*a*) (i) State the value of the p.d. across the capacitor when it is fully charged.

(ii) Calculate the maximum current during the charging process.

(iii) Sketch a graph showing how the p.d. across resistor R varies with time during the charging process. Numerical values are not required. **4**

(*b*) The student deduces from the graph of current against time for the discharge that the resistance of the lamp is less than 800 Ω.

Explain why the student's deduction is correct. **1**

(*c*) Calculate the energy stored in the capacitor when it is fully charged. **2**

(7)

Marks

8. (*a*) An operational amplifier is connected in a circuit as shown below.

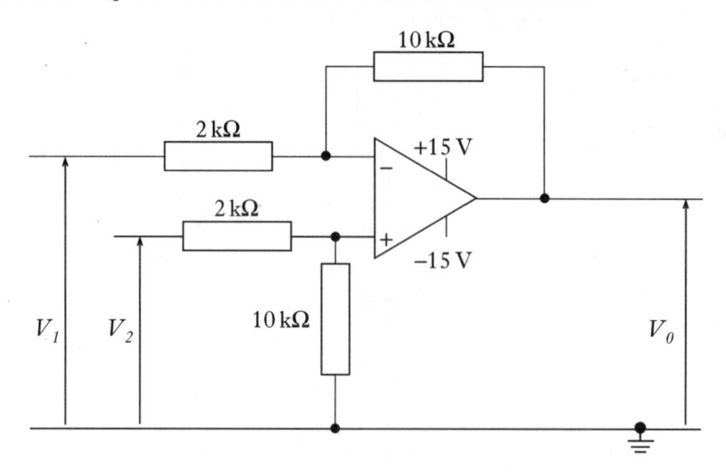

(i) In what mode is the operational amplifier operating?

(ii) The input voltage V_1 is 0·3 V and input voltage V_2 is 0·4 V.
Calculate the output voltage V_0. **3**

(*b*) A second operational amplifier is now connected as shown below.

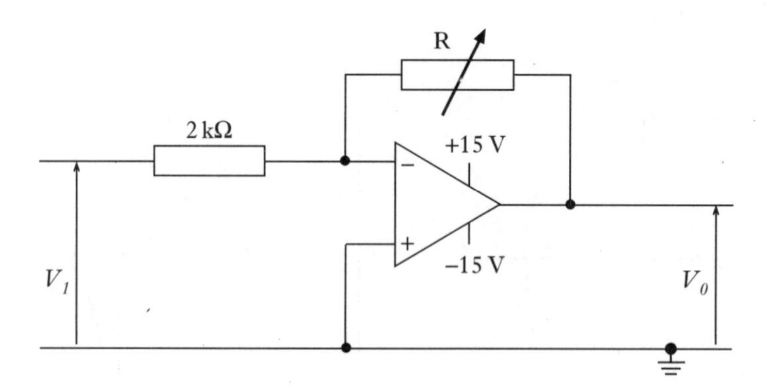

(i) The input voltage V_1 is 0·5 V and the output voltage V_0 is −4·0 V.
Calculate the resistance of R.

(ii) The input voltage V_1 is kept at 0·5 V. The resistance of R is gradually increased to 100 kΩ.
Describe what happens to the output voltage V_0. **4**

(7)

Marks

9. (*a*) The word Laser is an acronym for "light amplification by the stimulated emission of radiation".

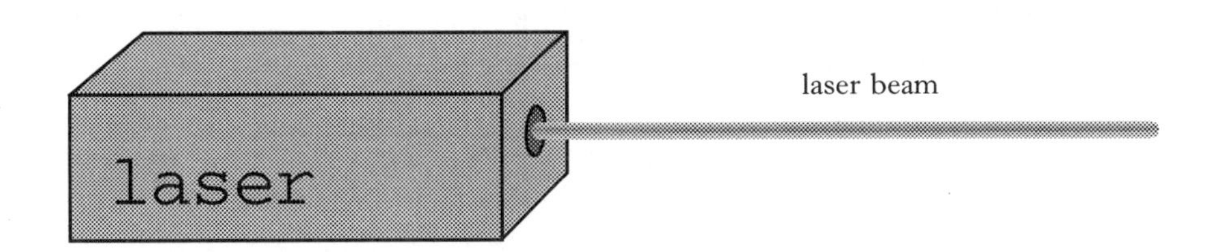

Describe what is meant by "stimulated emission" and describe how amplification is produced in a laser. **3**

(*b*) Infrared radiation from a laser is directed at a small solid cylinder of copper as shown below. The cylinder has a cross sectional area of $1{\cdot}25 \times 10^{-5}\,\text{m}^2$. The intensity of the laser beam at the surface of the cylinder is $4{\cdot}00 \times 10^{5}\,\text{W m}^{-2}$.

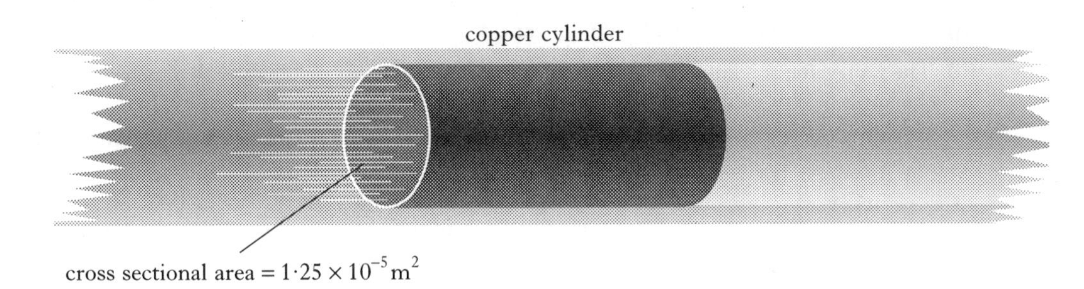

(i) Show that the energy delivered to the cylinder in 100 seconds is 500 J.

(ii) The cylinder has a mass of $1{\cdot}12 \times 10^{-3}\,\text{kg}$ and the initial temperature of the cylinder is 293 K.

Using information from the Data Sheet, show by calculation whether or not the 500 J of energy is sufficient to raise the temperature of the copper cylinder to its melting point. **7**

(10)

Marks

10. In investigating the effect of different types of radiation on the human body, the data in the table below was obtained for one particular type of body tissue.

Radiation	*Absorbed Dose Rate*	*Quality Factor*
γ-rays	$100\,\mu Gy\,h^{-1}$	1
Fast neutrons	$400\,\mu Gy\,h^{-1}$	10
α-particles	$50\,\mu Gy\,h^{-1}$	20

(*a*) Show, using the data in the table, which radiation is likely to be the most harmful to this tissue. **3**

(*b*) (i) The maximum permitted dose equivalent for this tissue is 5 mSv. Calculate the time the tissue can be exposed to fast neutrons without exceeding this limit.

(ii) A sample of this tissue has a mass of 25 grams. How much energy will it absorb from fast neutrons in 2 hours? **5**

(*c*) The effect of radiation on tissue can be reduced by putting shielding material between the source of radiation and the tissue. The effectiveness of this shielding material can be described by the half-value thickness of the material.

(i) Explain the meaning of "half-value thickness".

(ii) The half-value thickness for a particular material is 7 mm. A block of this material of thickness 3·5 cm is inserted between the source and the tissue.

What fraction of the radiation which is directed at the tissue is received by the tissue? **3**

(11)

Marks

11. (*a*) Two possible nuclear reactions involving uranium are represented by the statements shown below.

Statement **A** $^{235}_{92}U + ^{1}_{0}n = ^{134}_{52}Te + ^{98}_{40}Zr + 4^{1}_{0}n$

Statement **B** $^{235}_{92}U + ^{1}_{0}n = ^{144}_{56}Ba + ^{90}_{36}Kr + 2^{1}_{0}n$

The masses of the nuclei and particles involved in the reactions are as follows.

	Mass
$^{235}_{92}U$	$3{\cdot}901 \times 10^{-25}$ kg
$^{134}_{52}Te$	$2{\cdot}221 \times 10^{-25}$ kg
$^{98}_{40}Zr$	$1{\cdot}626 \times 10^{-25}$ kg
$^{144}_{56}Ba$	$2{\cdot}388 \times 10^{-25}$ kg
$^{90}_{36}Kr$	$1{\cdot}492 \times 10^{-25}$ kg
$^{1}_{0}n$	$0{\cdot}017 \times 10^{-25}$ kg

(i) What type of nuclear reaction is described by statements **A** and **B**?

(ii) Show by calculation how much mass is "lost" in each of reactions **A** and **B**.

(iii) Explain which of the reactions **A** and **B** releases the greater amount of energy. **6**

(*b*) A third possible nuclear reaction involving $^{235}_{92}U$ is represented by the following statement.

$$^{235}_{92}U + ^{1}_{0}n = ^{98}_{42}Mo + ^{136}_{y}Xe + 2^{1}_{0}n + 4^{0}_{-1}e$$

(i) The symbol for the uranium nucleus is $^{235}_{92}U$. What information about the particles in the nucleus is provided by the numbers 92 and 235?

(ii) Determine the number represented by y. **3**

(9)

[*END OF QUESTION PAPER*]

SOLUTIONS — PAPER II

Fully worked solutions to these questions are given in our book entitled *Solutions to Higher Grade Physics*

SCOTTISH CERTIFICATE OF EDUCATION 1997

THURSDAY, 15 MAY
9.30 AM – 11.00 AM

PHYSICS
HIGHER GRADE
Paper I

Read Carefully

1 All questions should be attempted.

2 The following data should be used when required unless otherwise stated.

Speed of light in vacuum c	$3{\cdot}00 \times 10^{8}$ m s^{-1}	Planck's constant h	$6{\cdot}63 \times 10^{-34}$ J s
Charge on electron e	$-1{\cdot}60 \times 10^{-19}$ C	Mass of electron m_e	$9{\cdot}11 \times 10^{-31}$ kg
Acceleration due to gravity g	$9{\cdot}8$ m s^{-2}	Mass of proton m_p	$1{\cdot}67 \times 10^{-27}$ kg

Section A (questions 1 to 30)

3 Check that the answer sheet is for Physics Higher I (Section A).

4 Answer the questions numbered 1 to 30 on the answer sheet provided.

5 Fill in the details required on the answer sheet.

6 Rough working, if required, should be done only on this question paper, or on the first two pages of the answer book provided—**not** on the answer sheet.

7 For each of the questions 1 to 30 there is only **one** correct answer and each is worth 1 mark.

8 Instructions as to how to record your answers to questions 1–30 are previously given.

Section B (questions 31 to 38)

9 Answer questions numbered 31 to 38 in the answer book provided.

10 Fill in the details on the front of the answer book.

11 Enter the question number clearly in the margin of the answer book beside each of your answers to questions 31 to 38.

12 Care should be taken **not** to give an unreasonable number of significant figures in the final answers to calculations.

1997

SECTION A

Answer questions 1–30 on the answer sheet.

1. Which of the following groups contains two vector quantities and one scalar quantity?

A Time, distance and force

B Acceleration, mass and momentum

C Velocity, force and momentum

D Displacement, velocity and acceleration

E Speed, distance and momentum

2. The diagram below is the velocity-time graph for a model train moving along a straight track.

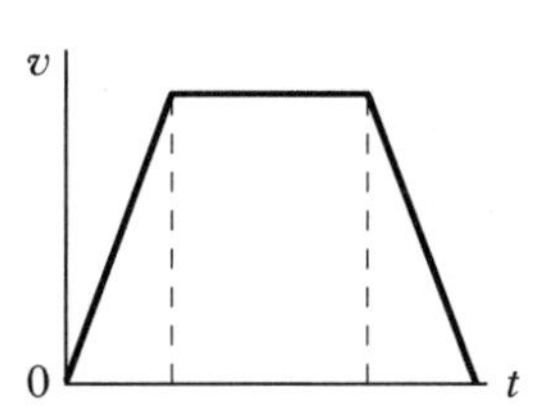

Which of the following could represent the displacement-time graph for the same motion?

A
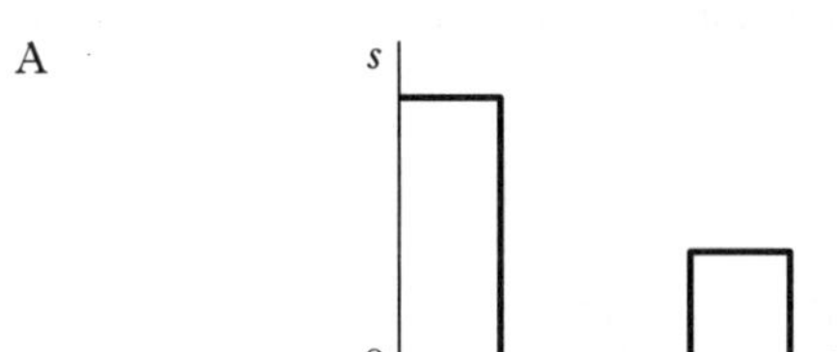

B

C
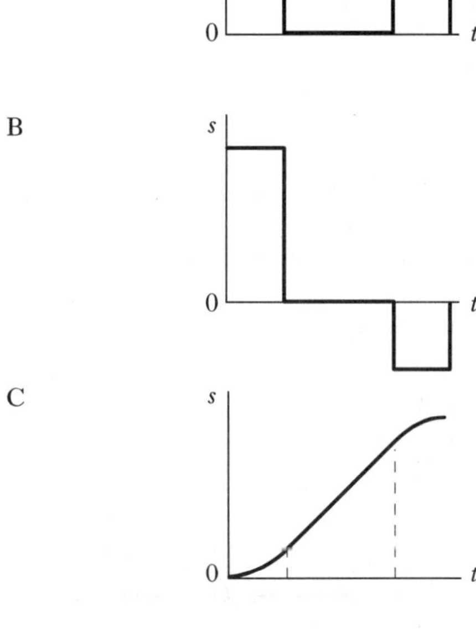

D
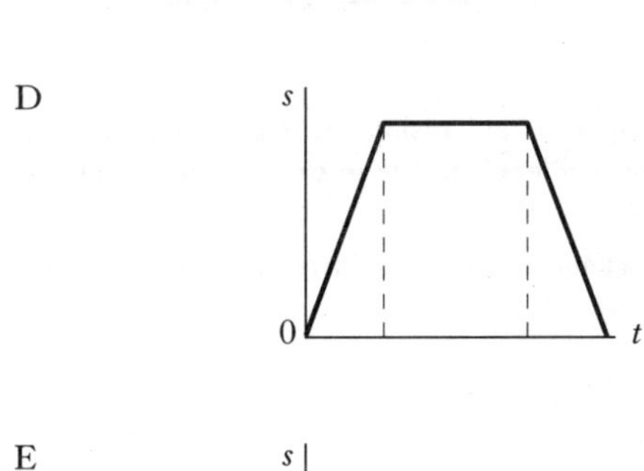

E
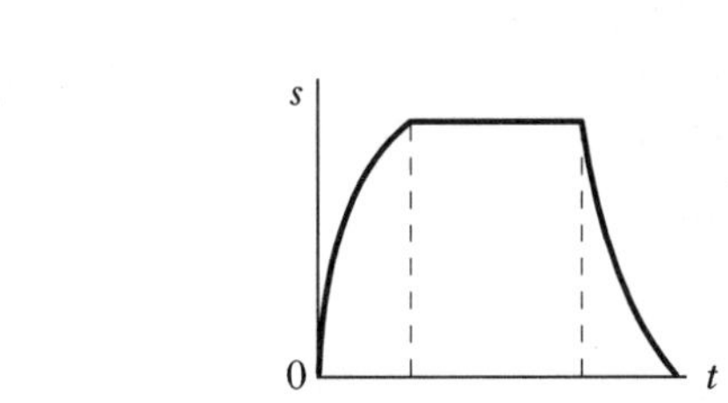

3. A train decelerates uniformly from $12{\cdot}0\ \mathrm{m\,s^{-1}}$ to $5{\cdot}0\ \mathrm{m\,s^{-1}}$ while travelling a distance of $119{\cdot}0$ m along a straight track. The deceleration of the train is

A $0{\cdot}5\ \mathrm{m\,s^{-2}}$

B $0{\cdot}7\ \mathrm{m\,s^{-2}}$

C $1{\cdot}2\ \mathrm{m\,s^{-2}}$

D $7{\cdot}0\ \mathrm{m\,s^{-2}}$

E $14{\cdot}0\ \mathrm{m\,s^{-2}}$.

4. A ball is projected vertically upwards with an initial speed of $40\ \mathrm{m\,s^{-1}}$. The acceleration due to gravity can be taken to be $10\ \mathrm{m\,s^{-2}}$.

What total time will the ball take to rise to its highest point and then return to its starting point?

A 2 s

B 4 s

C 6 s

D 8 s

E 16 s

5. An aeroplane is flying at $160\ \mathrm{m\,s^{-1}}$ in level flight 80 m above the ground. It releases a package at a horizontal distance X from the target T.

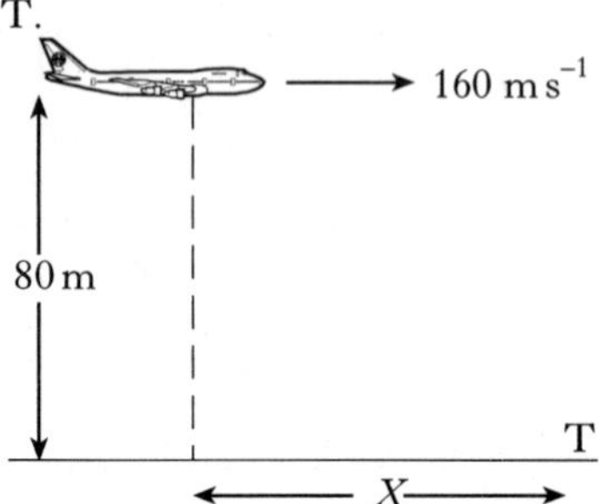

The effect of air resistance can be neglected and the acceleration due to gravity can be taken as $10\ \mathrm{m\,s^{-2}}$.

The package will score a direct hit on the target T if X is

A 40 m

B 160 m

C 320 m

D 640 m

E 2560 m.

6. The lift in a department store has a mass of 1100 kg.

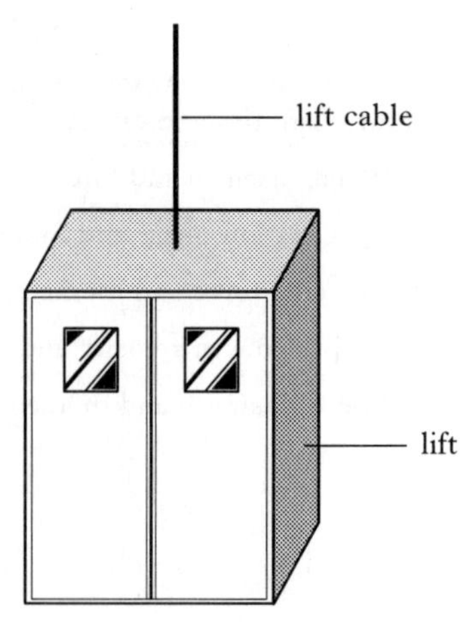

The lift is descending with a uniform downwards acceleration of $2\ \mathrm{m\,s^{-2}}$. The acceleration due to gravity can be taken as $10\ \mathrm{m\,s^{-2}}$.

What is the force applied to the lift by the lift cable?

A 1100 N

B 2200 N

C 8800 N

D 11 000 N

E 13 200 N

7. A sledge is pulled a distance of 8 m in a straight line along a horizontal surface.

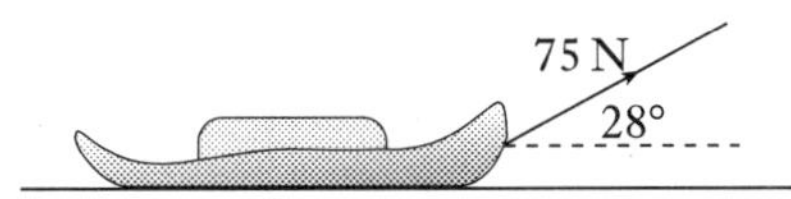

The tension in the rope is 75 N and the angle between the rope and the horizontal surface is 28°.

Which row in the following table is correct?

	Horizontal component of tension/N	*Vertical component of tension*/N	*Work done by rope*/J
A	75 sin 28°	75 sin 62°	600
B	75 cos 28°	75 sin 28°	530
C	75 sin 62°	75 sin 28°	600
D	75 cos 28°	75 sin 62°	600
E	75 sin 28°	75 cos 28°	35

8. Many car manufacturers are now fitting airbags which inflate automatically during an accident, as shown below.

The purpose of the airbag is to protect the driver by

A reducing his change of momentum per second

B increasing his change of momentum per second

C reducing his final velocity

D reducing his total change in momentum

E increasing his total change in momentum.

9. The force acting on an object is measured and the results are stored in a computer.

The force-time graph obtained from the computer is shown below.

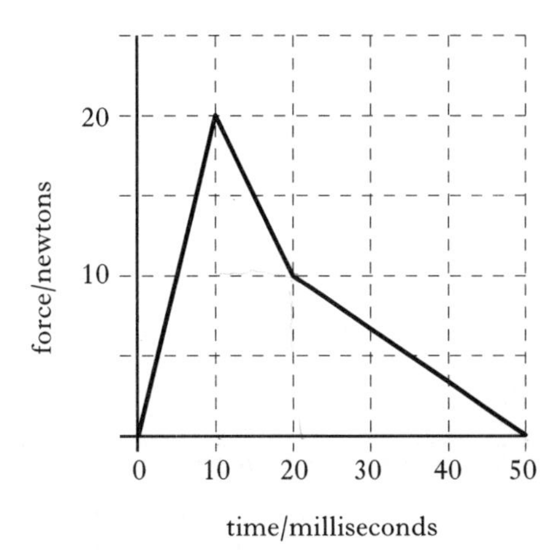

What is the average force acting on the object during the 50 milliseconds?

A 15 N

B 10 N

C 8 N

D 2·5 N

E 1 N

10. A small metal block is suspended from a spring balance at a depth h below the surface of a liquid in a large beaker.

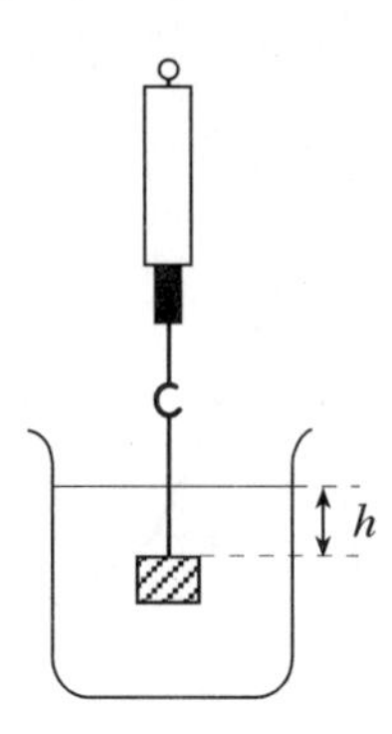

Which of the following statements is/are true?

I The reading on the spring balance depends on the density of the liquid in the beaker.

II The reading on the spring balance is equal to the upthrust of the liquid on the metal block.

III The reading on the spring balance will increase as the depth h is increased.

A I only

B II only

C III only

D I and II only

E I and III only

11. Which of the following gives the approximate relative spacings of molecules in ice, water and water vapour?

	Molecular spacing in ice/units	*Molecular spacing in water/units*	*Molecular spacing in water vapour/units*
A	1	1	10
B	1	3	1
C	1	3	3
D	1	10	10
E	3	1	10

12. Two parallel metal plates X and Y in a vacuum have a potential difference V across them.

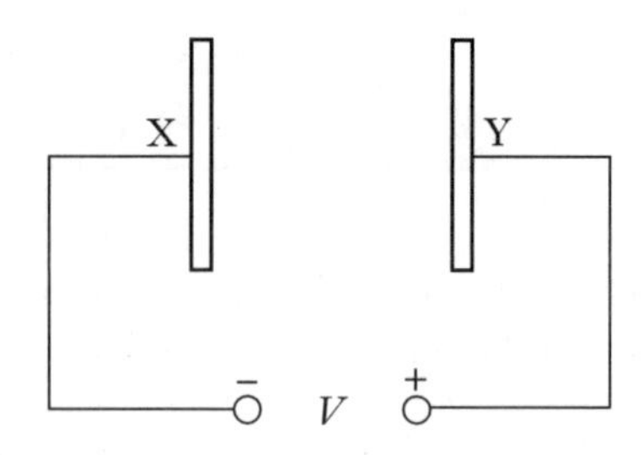

An electron of charge e and mass m, initially at rest, is released from plate X.

The speed of the electron when it reaches plate Y is given by

A $\dfrac{2eV}{m}$

B $\sqrt{\dfrac{2eV}{m}}$

C $\sqrt{\dfrac{2V}{em}}$

D $\dfrac{2V}{em}$

E $\dfrac{2mV}{e}$

13. In the following circuit, the p.d. across the $16\,\Omega$ resistor is 40 V when switch S is **open**.

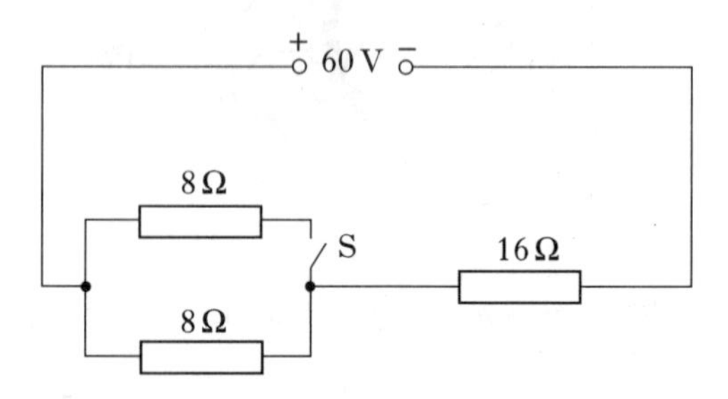

The p.d. across the $16\,\Omega$ resistor when switch S is **closed** is

A 12 V

B 15 V

C 30 V

D 45 V

E 48 V.

14. In the circuit shown, the p.d. between points P and Q is 12 V.

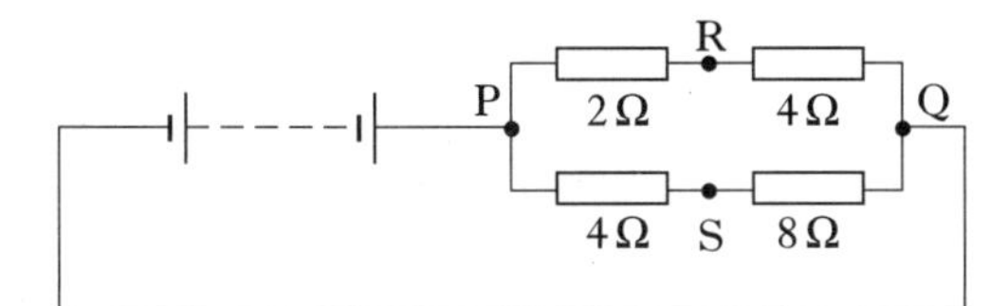

The reading on a voltmeter connected across points R and S is

A 0 V

B 2 V

C 4 V

D 6 V

E 8 V.

15. When there is a potential difference of V volts across a resistor, the power dissipated in the resistor is P watts.

The current in the resistor, in amperes, is given by

A $\frac{P}{V}$

B $\frac{P}{V^2}$

C $\frac{V}{P}$

D $\frac{V^2}{P}$

E $\sqrt{\frac{P}{V}}$

16. The unit for capacitance can be written as

A $\mathrm{V\,C^{-1}}$

B $\mathrm{C\,V^{-1}}$

C $\mathrm{J\,s^{-1}}$

D $\mathrm{C\,J^{-1}}$

E $\mathrm{J\,C^{-1}}$.

17. The following circuit is used to charge and then discharge a capacitor C.

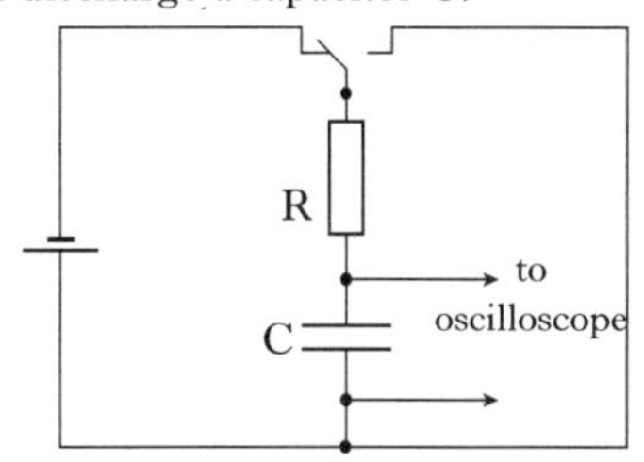

Which of the following pairs of graphs correctly shows how the voltage V across the capacitor varies with time during charging and discharging?

Charging **Discharging**

A V 0 time V 0 time

B V 0 time V 0 time

C
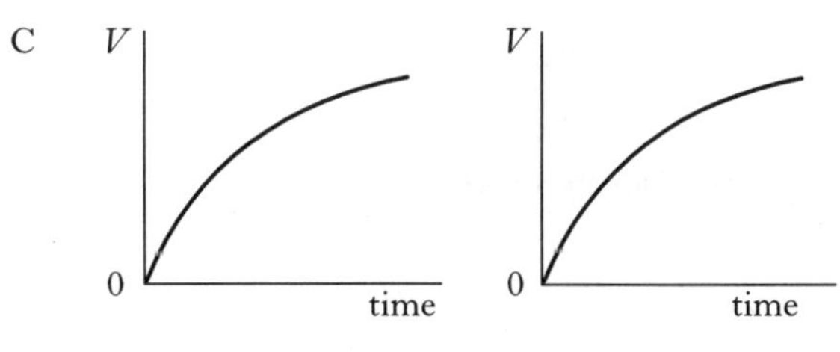

D
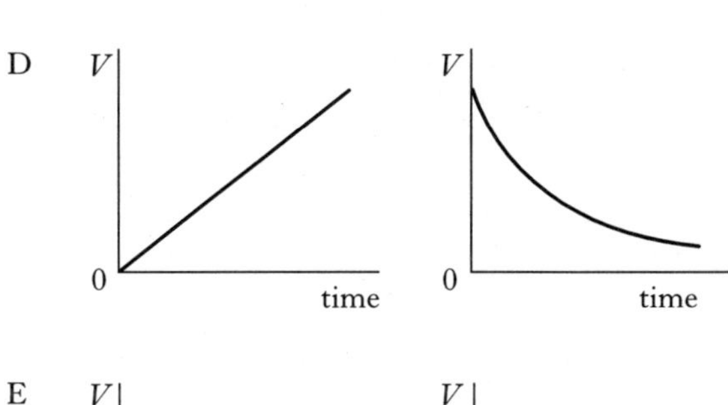

E
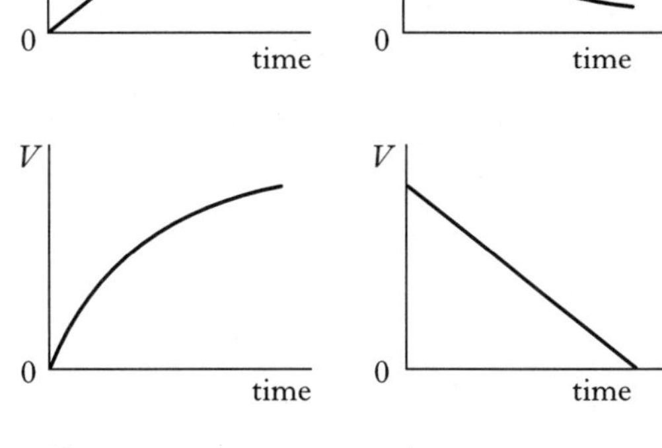

18. A capacitor is to be connected across a 230 V r.m.s. a.c. supply. To prevent damage to the capacitor, its minimum voltage rating must be

A 163 V

B 230 V

C 325 V

D 460 V

E 650 V.

19. A resistor is connected in a circuit as shown below.

The output from the a.c. supply can be varied in frequency but has a constant r.m.s. voltage.

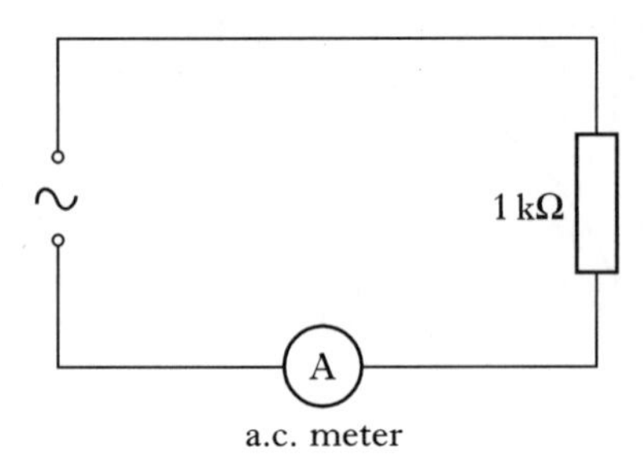

The frequency of the output from the supply is increased steadily from 50 Hz to 5 kHz.

The reading on the a.c. ammeter

A remains constant

B falls steadily

C rises steadily

D rises then falls

E falls then rises.

20. A physicist designs the amplifier circuit shown below.

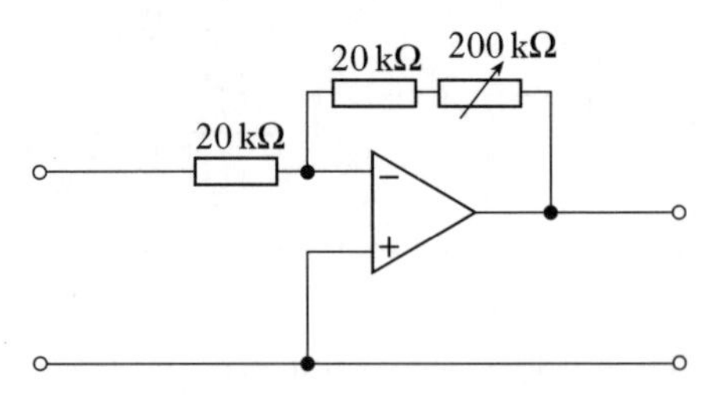

In this circuit, adjustment of the resistance of the variable resistor from zero to 200 kΩ allows the voltage gain to be altered over the range

A zero to one

B zero to ten

C zero to eleven

D one to ten

E one to eleven.

21. A source of microwaves of wavelength λ is placed behind two slits, R and S. A microwave detector records the maximum response when it is placed at P, where RP = SP.

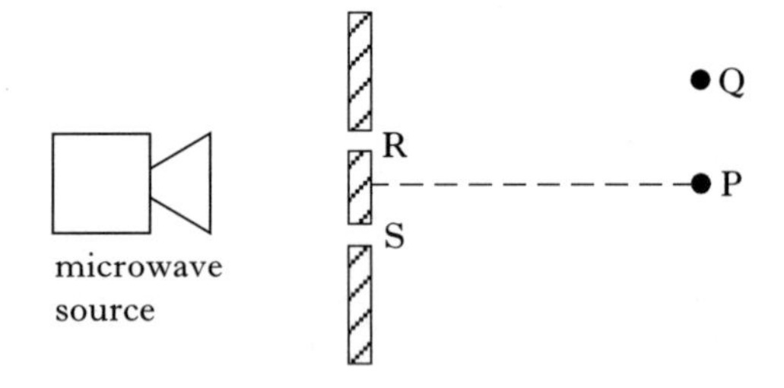

The microwave detector is moved and the **next** maximum is recorded at Q.

The path difference (SQ–RQ) must be

A 0

B $\frac{\lambda}{2}$

C λ

D (any odd number) $\times \frac{\lambda}{2}$

E (any whole number) $\times \lambda$

22. A liquid and a solid have the same refractive index.

What would happen to the speed and the wavelength of light waves passing from the liquid into the solid?

	Speed	*Wavelength*
A	decreases	decreases
B	decreases	increases
C	increases	increases
D	increases	decreases
E	stays the same	stays the same

23. The intensity of light can be measured in

A W

B $W m^{-1}$

C $W m$

D $W m^{-2}$

E $W m^{2}$.

24. The light intensity is 160 units at a distance of 0·50 m from a point source of light in a darkened room.

At 2·0 m from this source, the light intensity is

A 160 units

B 80 units

C 40 units

D 10 units

E 5 units.

25. The diagram below shows some of the energy levels for the hydrogen atom.

E_3 ———— $-1{\cdot}360 \times 10^{-19}$ J

E_2 ———— $-2{\cdot}416 \times 10^{-19}$ J

E_1 ———— $-5{\cdot}424 \times 10^{-19}$ J

E_0 ———— $-21{\cdot}76 \times 10^{-19}$ J

The highest frequency of radiation emitted due to a transition between two of these energy levels is

A $2{\cdot}04 \times 10^{20}$ Hz

B $1{\cdot}63 \times 10^{20}$ Hz

C $3{\cdot}08 \times 10^{15}$ Hz

D $2{\cdot}46 \times 10^{15}$ Hz

E $1{\cdot}59 \times 10^{14}$ Hz.

26. A light emitting diode produces light of wavelength λ.

The energy of a photon of light emitted by this diode is given by

A $h\lambda$

B $\dfrac{h}{\lambda}$

C $\dfrac{h\lambda}{c}$

D $\dfrac{hc}{\lambda}$

E $hc\lambda$

27. Which of the following statements could explain the faint dark lines observed in the spectrum of sunlight when viewed through a high quality spectroscope?

I Gases in the outer layers of the Sun absorb certain frequencies of light.

II Gases in the inner layers of the Sun emit only certain frequencies of light.

III Gases within the Sun produce only a line emission spectrum.

A I only

B II only

C III only

D I and II only

E I and III only

28. A physicist varied the distance between a radioactive source and a detector. She obtained the following results which have been corrected for background radiation.

Distance from source to detector/cm	10	20	30	40	50
Corrected count-rate/s^{-1}	127	31	14	8	5

From these results, what is the relationship between the corrected count-rate R and the distance d from the source to the detector?

A $R \propto d^2$

B $R \propto d$

C $R \propto \sqrt{d}$

D $R \propto \frac{1}{d}$

E $R \propto \frac{1}{d^2}$

29. Part of a radioactive decay series is shown below.

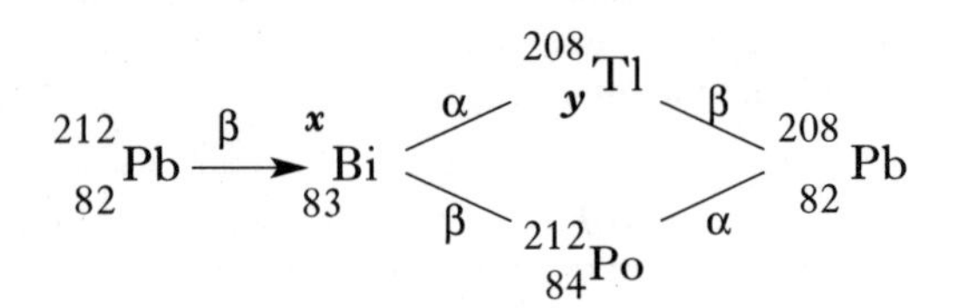

The numbers x and y in the series have been omitted.
What are the correct values for x and y?

	x	y
A	212	84
B	211	81
C	213	84
D	212	81
E	211	83

30. The process of nuclear fission occurs in the core of a nuclear reactor.

Which of the following statements about this process is/are true?

I Two nuclei are produced when an unstable nucleus fissions.

II Two unstable nuclei combine.

III Neutrons are released when an unstable nucleus fissions.

A I only

B II only

C III only

D I and II only

E I and III only

SECTION B

Write your answers to questions 31 to 38 in the answer book.

Marks

31. Two ropes are used to pull a boat at constant speed along a canal.

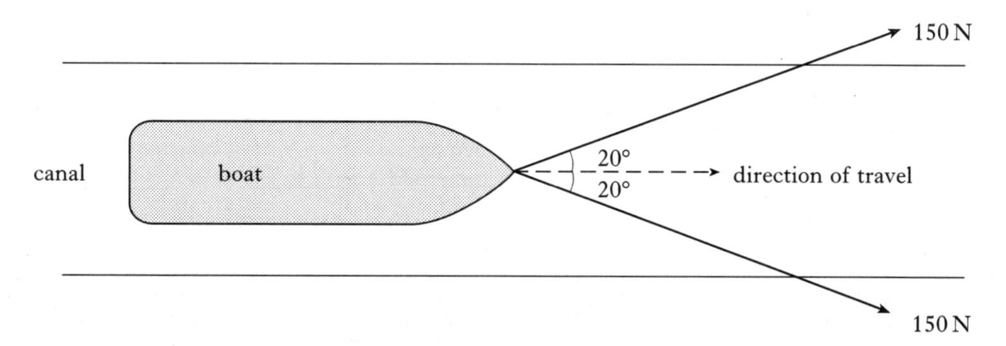

Each rope exerts a force of 150 N at 20° to the direction of travel of the boat as shown.

(*a*) Calculate the magnitude of the resultant force exerted by the ropes.

(*b*) What is the magnitude of the frictional force acting on the boat? **3**

32. The diagram shows a weather balloon of mass *m* tethered by a rope to the ground.

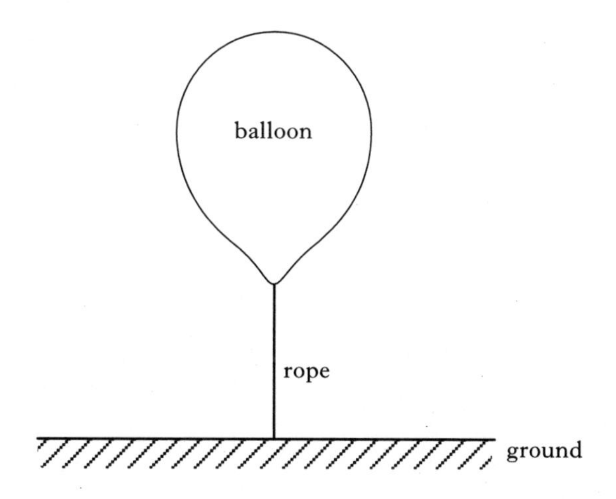

(*a*) Draw a sketch of the balloon. Mark and name all the forces acting vertically on the balloon.

(*b*) What is the resultant force acting on the balloon? **2**

Marks

33. A pupil carries out an experiment on a linear air track with two vehicles X and Y. Vehicle X is propelled towards vehicle Y which is initially at rest and the vehicles are allowed to collide.

The results obtained are shown in the tables below.

Before Collision			
Momentum of X/ kg m s^{-1}	*Momentum of Y*/ kg m s^{-1}	*Kinetic energy of X*/J	*Kinetic energy of Y*/J
0·12	0	0·036	0

After Collision			
Momentum of X/ kg m s^{-1}	*Momentum of Y*/ kg m s^{-1}	*Kinetic energy of X*/J	*Kinetic energy of Y*/J
0·06	0·06	0·009	0·018

Explain whether the collision between the vehicles is elastic or inelastic. **2**

34. The output from a signal generator is connected to the input terminals of an oscilloscope. A trace is obtained on the oscilloscope screen. The oscilloscope control settings and the trace on the oscilloscope screen are shown in the diagram below.

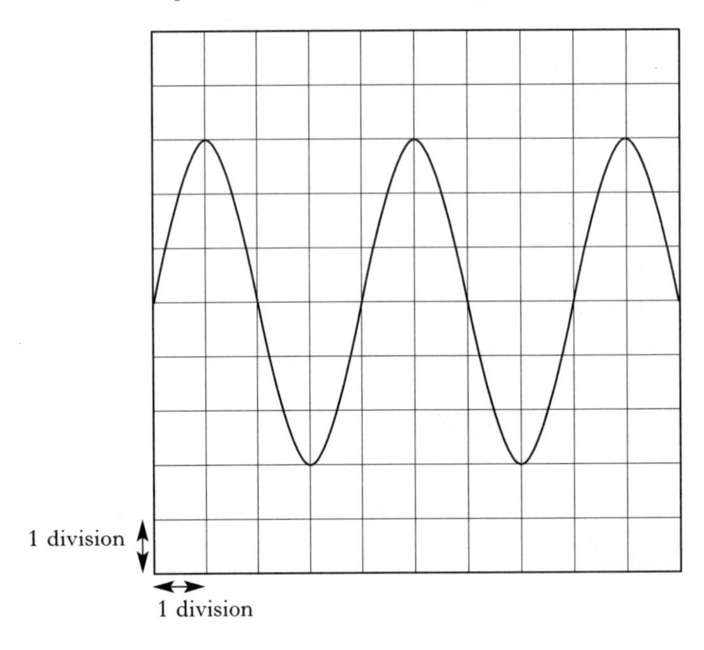

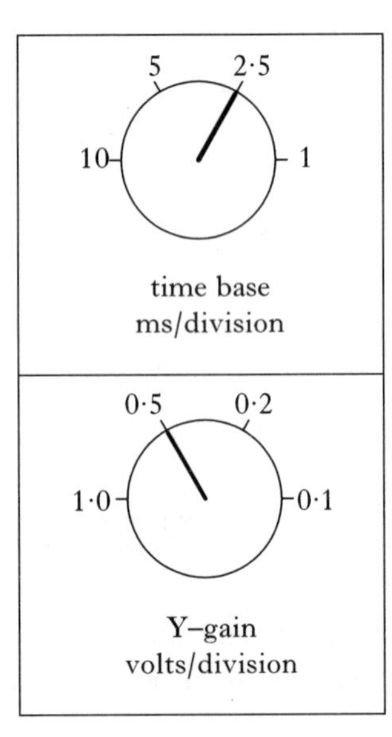

(*a*) Calculate the frequency of the output from the signal generator.

(*b*) The frequency and amplitude of the output from the signal generator are kept constant.

The time base control setting is changed to 5 ms/division.

What will be the effect on the trace shown on the oscilloscope? **3**

Marks

35. The graph shows how the voltage across the terminals of a battery changes as the current from the battery is varied.

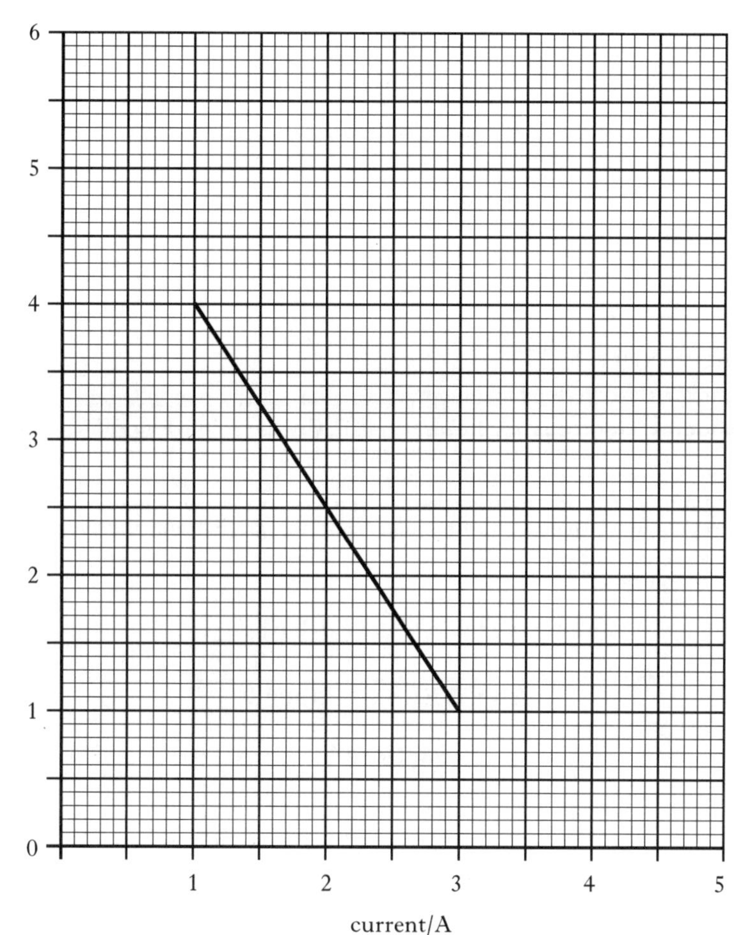

(*a*) Calculate the internal resistance of the battery.

(*b*) What is the value of the current from the battery when it is short-circuited? **3**

36. A beam of monochromatic light of frequency $4{\cdot}85 \times 10^{14}$ Hz passes from air into liquid paraffin. In liquid paraffin the light has a speed of $2{\cdot}10 \times 10^{8}$ m s^{-1}.

(*a*) Calculate the refractive index of the liquid paraffin.

(*b*) What is the frequency of the light when it is in the liquid paraffin? **3**

Marks

37. The diagram shows a photodiode connected to a voltmeter. A lamp is used to shine light onto the photodiode.

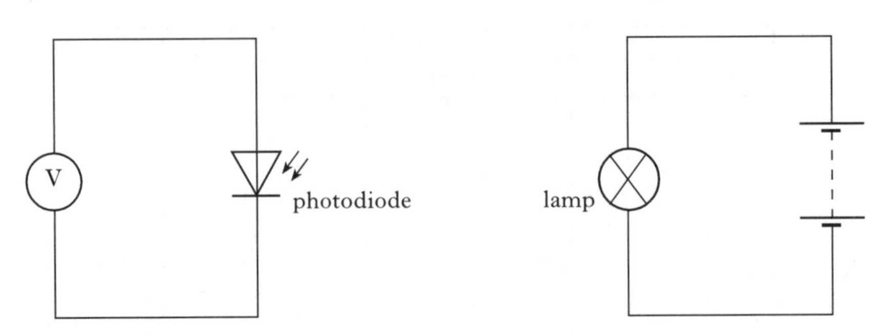

The reading on the voltmeter is 0·5 V.

The lamp is now moved closer to the photodiode.

Using the terms **photons**, **electrons** and **holes**, explain why the voltmeter reading changes. **2**

38. A grating or a prism can be used to produce spectra from a source of white light.
Give **two** differences between the spectra obtained using the grating and the prism. Diagrams may be used to illustrate your answer. **2**

[*END OF QUESTION PAPER*]

ANSWERS — Paper I

Section A

1. B	**2.** C	**3.** A	**4.** D	**5.** D	**6.** C	**7.** B	**8.** A	**9.** C	**10.**
11. A	**12.** B	**13.** E	**14.** A	**15.** A	**16.** B	**17.** A	**18.** C	**19.** A	**20.**
21. C	**22.** E	**23.** D	**24.** D	**25.** C	**26.** D	**27.** A	**28.** E	**29.** D	**30.**

Section B

31. *(a)* 282 N *(b)* 282 N

32. *(a)*

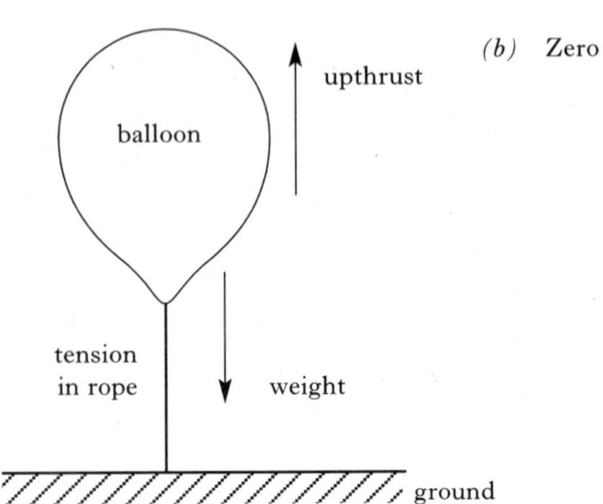

(b) Zero

33. E_k before = 0·036 J
E_k after = 0·027 J
Hence inelastic collision

34. *(a)* $f = 100$ Hz
(b) Trace half as wide i.e. 5 waves on screen but same height

35. *(a)* $r = 1{\cdot}5\ \Omega$ *(b)* $I = 3{\cdot}67$ A

36. *(a)* $n_p = 1{\cdot}428$ *(b)* $4{\cdot}85 \times 10^{14}$ Hz

37. More **photons** from lamp now enter photodiode. More energy liberates more **electrons** at the junction creating m **electron/hole** pairs, creating more **voltage**.

38. Prism, one spectrum.
Grating many.
Order of colours reversed for grating.

SCOTTISH CERTIFICATE OF EDUCATION 1997

THURSDAY, 15 MAY
1.00 PM – 3.30 PM

PHYSICS
HIGHER GRADE
Paper II

Read carefully

1 All questions should be attempted.

2 Enter the question number clearly in the margin beside each question.

3 Any necessary data will be found in the Data Sheet.

4 Care should be taken not to give an unreasonable number of significant figures in the final answers to calculations.

5 Square-ruled paper (if used) should be placed inside the front cover of the answer book for return to the Scottish Qualifications Authority.

Marks

1. (*a*) An object starts from rest and moves with constant acceleration a. After a time t, the velocity v and displacement s are given by

$v = at$ and $s = \frac{1}{2}at^2$ respectively.

Use these relationships, to show that

$v^2 = 2as$. **2**

(*b*) An aircraft of mass of 1000 kg has to reach a speed of 33 $m\,s^{-1}$ before it takes off from a runway. The engine of the aircraft provides a constant thrust of 3150 N. A constant frictional force of 450 N acts on the aircraft as it moves along the runway.

(i) Calculate the acceleration of the aircraft along the runway.

(ii) The aircraft starts from rest. What is the minimum length of runway required for a take-off? **4**

(*c*) During a flight the aircraft is travelling with a velocity of 36 $m\,s^{-1}$ due north (000).

A wind with a speed of 12 $m\,s^{-1}$ starts to blow **towards** the direction 40° west of north (320).

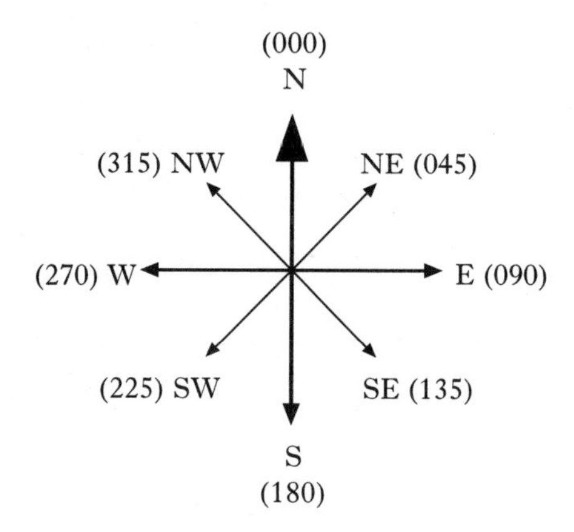

Find the magnitude and direction of the resultant velocity of the aircraft. **3**

(9)

Marks

2. The fairway on a golf course is in two horizontal parts separated by a steep bank as shown below.

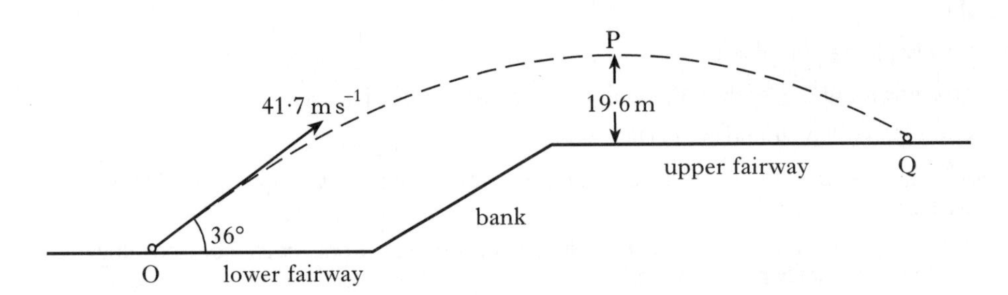

A golf ball at point O is given an initial velocity of 41·7 m s^{-1} at 36° to the horizontal.

The ball reaches a maximum vertical height at point P above the upper fairway. Point P is 19·6 m above the upper fairway as shown. The ball hits the ground at point Q.

The effect of air friction on the ball may be neglected.

(*a*) Calculate:

(i) the horizontal component of the initial velocity of the ball;

(ii) the vertical component of the initial velocity of the ball. **2**

(*b*) Show that the time taken for the ball to travel from point O to point Q is 4·5 s. **3**

(*c*) Calculate the horizontal distance travelled by the ball. **2**

(7)

3. The diagram below shows two vehicles P and Q on a linear air track.

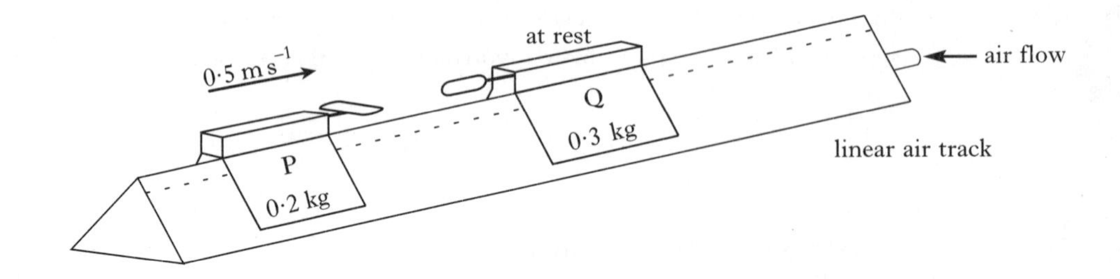

Vehicle P, of mass 0·2 kg, is projected with a velocity of 0·5 m s^{-1} to the right along the linear air track.

It collides with vehicle Q, of mass 0·3 kg, which is initially at rest.

After the collision, the vehicles move in opposite directions. Vehicle Q moves off with a velocity of 0·4 m s^{-1} to the right.

(*a*) Show that vehicle P rebounds with a speed of 0·1 m s^{-1} after the collision. **2**

(*b*) Calculate the change in momentum of vehicle P as a result of the collision. **2**

(*c*) During the collision, a timing device records the time of contact between the two vehicles as 0·06 s.

(i) Calculate the average force acting on vehicle P during the collision.

(ii) Sketch a graph showing how the force on vehicle P could vary with time while the two vehicles are in contact. **3**

(7)

Marks

4. A pupil uses the apparatus shown in the diagram to investigate the relationship between the pressure and the temperature of a fixed mass of gas at constant volume.

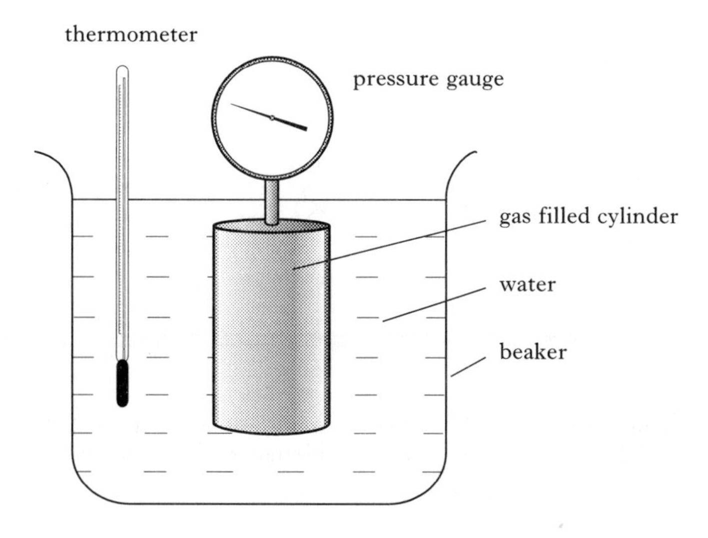

The cylinder is fully immersed in a beaker of water and the water is slowly heated.

You may assume that the volume of the cylinder does not change as the temperature of the water changes.

(*a*) Explain why the cylinder must be fully immersed in the beaker of water. **1**

(*b*) The pressure of the gas in the cylinder is 100 kPa when the gas is at a temperature of 17 °C.
Calculate the pressure of the gas in the cylinder when the temperature of the gas is 75 °C. **2**

(*c*) The base of the cylinder has an area of $0{\cdot}001\ m^2$.
What is the force exerted by the gas on the base when the temperature of the gas is 75 °C? **2**

(*d*) What happens to the density of the gas in the cylinder as the temperature increases from 17 °C to 75 °C?
Justify your answer. **2**

(7)

Marks

5. The diagram shows a circuit for part of the electrical system of a car.

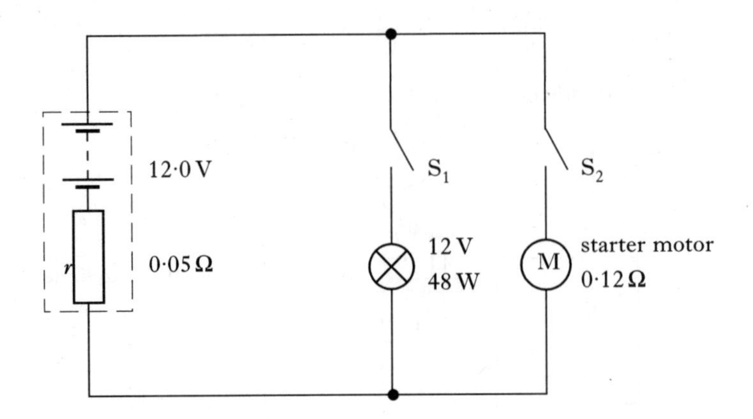

The battery has an e.m.f. of 12·0 V and an internal resistance r of 0·05 Ω. The battery is connected across a 12 V, 48 W headlamp and a starter motor of resistance 0·12 Ω as shown.

(*a*) State what is meant by "the battery has an e.m.f. of 12·0 V". **1**

(*b*) (i) What is the resistance of the headlamp when used at its rated voltage?

(ii) Show that there is a p.d. of 11·8 V across the headlamp when switch S_1 is closed and switch S_2 is open. Assume that the resistance of the headlamp does not change. **4**

(*c*) Both switches S_1 and S_2 are now closed.

Assuming that the resistance of the headlamp does not change, calculate:

(i) the total resistance of the circuit;

(ii) the current from the battery. **4**

(9)

Marks

6. (*a*) An operational amplifier circuit is set up with oscilloscopes connected across PQ and XY as shown.

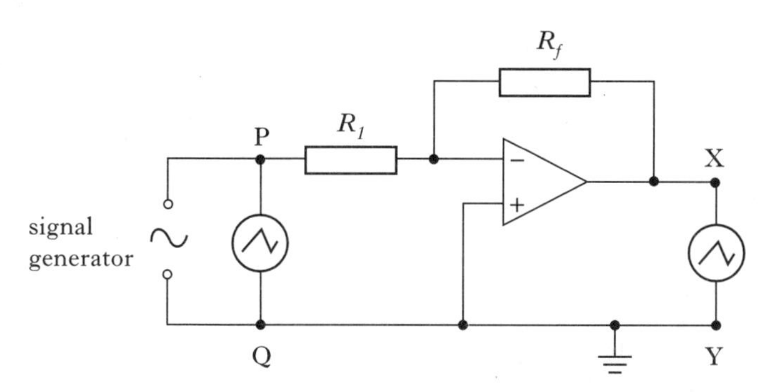

The trace on the oscilloscope connected across PQ is shown in figure 1 below. The Y–gain setting of this oscilloscope is 2 mV/division.

The trace on the oscilloscope connected across XY is shown in figure 2 below. The Y–gain setting of this oscilloscope is 30 mV/division.

The time base on each oscilloscope is set at 0·1 ms/division.

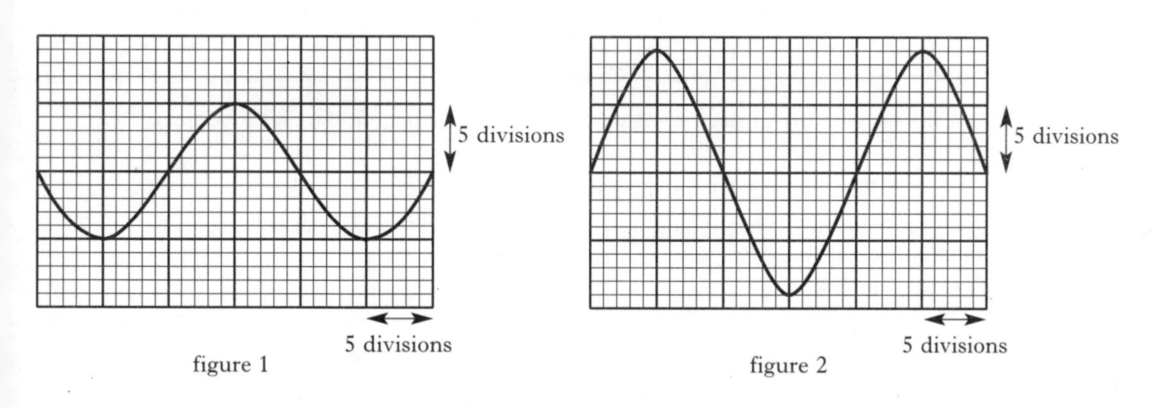

figure 1

figure 2

(i) Calculate the voltage gain of the amplifier.

(ii) Calculate the r.m.s. value of the output voltage from the amplifier.

(iii) Suggest suitable values for R_f and R_1 which would produce the trace shown in figure 2. **5**

Marks

(*b*) An electronic circuit is used to monitor temperature during an experiment. The circuit includes a Wheatstone bridge and an operational amplifier operating in the differential mode. One of the components of the Wheatstone bridge is a thermistor.

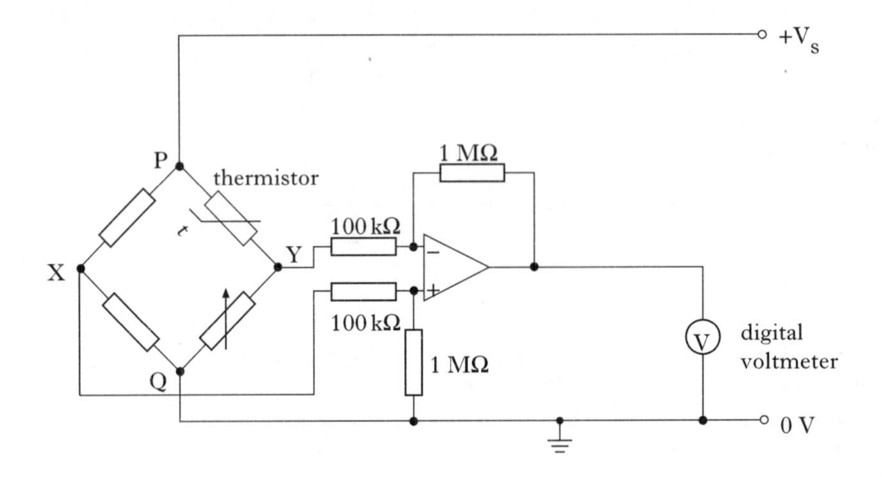

(i) When the temperature of the thermistor is 23 °C, the reading on the digital voltmeter is 0·00 V.

What is the value of the p.d. between X and Y?

(ii) When the temperature of the thermistor is 26 °C, the reading on the digital voltmeter is −0·18 V.

What is now the value of the p.d. between X and Y?

(iii) State what happens to the reading on the digital voltmeter when the temperature of the thermistor falls to 20 °C.

Justify your answer. **5**

(10)

Marks

7. (*a*) A capacitor of capacitance 220 μF is connected in series with a 150 kΩ resistor, a switch and an ammeter. A d.c. power supply of negligible internal resistance is connected to the circuit as shown below.

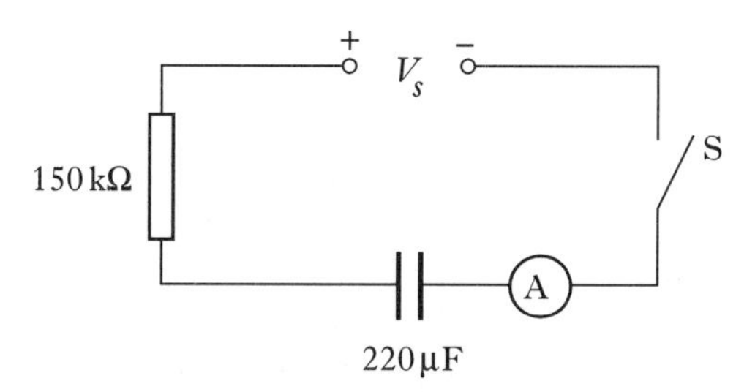

A stopclock is started and after 10 seconds the switch S is closed. Ammeter readings are noted at regular intervals until a time of 80 s is shown on the stopclock.

The graph below shows how the current in the circuit varies with time.

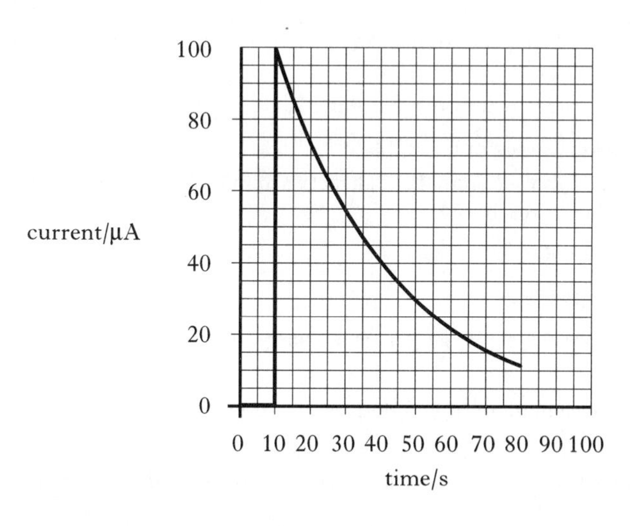

(i) Calculate the voltage V_s of the d.c. power supply.

(ii) At what time on the stopclock does the p.d. across the resistor equal 6 V?

(iii) What is the p.d. across the capacitor when the p.d. across the resistor is 6 V? **5**

(*b*) A magazine article on the resuscitation of a heart attack victim describes the equipment used. This equipment uses a 16 μF capacitor which is charged until the p.d. across it is 6 kV. The capacitor is then fully discharged to give the heart a shock. The discharge time is 2 ms.

(i) When the capacitor is fully charged, calculate:

(A) the charge stored;

(B) the energy stored.

(ii) Calculate the average current during discharge. **6**

(11)

Marks

8. Two identical loudspeakers X and Y are set up in a room which has been designed to eliminate the reflection of sound. The loudspeakers are connected to the same signal generator as shown.

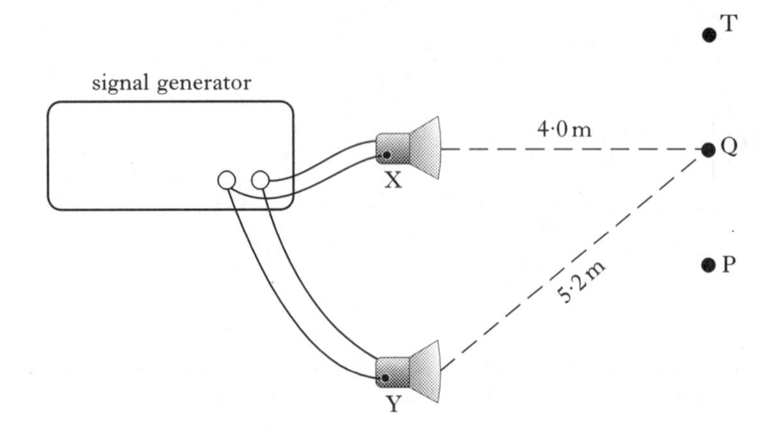

(*a*) (i) When a sound level meter is moved from P to T, maxima and minima of sound intensity are detected.

Explain, in terms of waves, why the maxima and minima are produced.

(ii) The sound level meter detects a maximum at P.

As the sound level meter is moved from P, it detects a minimum then a maximum then another minimum when it reaches Q.

Calculate the wavelength of the sound used. **4**

(*b*) The sound level meter is now fixed at Q.

The frequency of the output from the signal generator is increased steadily from 200 Hz to 1000 Hz.

(i) What happens to the wavelength of the sound as the frequency of the output is increased?

(ii) Explain why the sound level meter detects a series of maxima and minima as the frequency of the output is increased. **3**

(7)

Marks

9. (*a*) The diagram below shows the refraction of a ray of red light as it passes through a plastic prism.

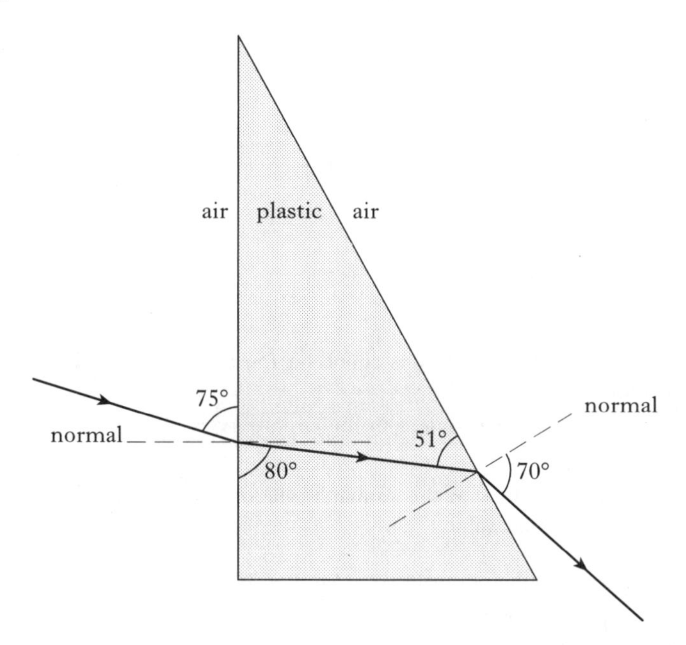

Calculate the refractive index of the plastic for this red light. **2**

(*b*) The refractive index of a glass block is found to be 1·44 when red light is used.

(i) What is the value of the critical angle for this red light in the glass?

(ii) The diagram shows the paths of two rays of this red light, PO and QO, in the glass block.

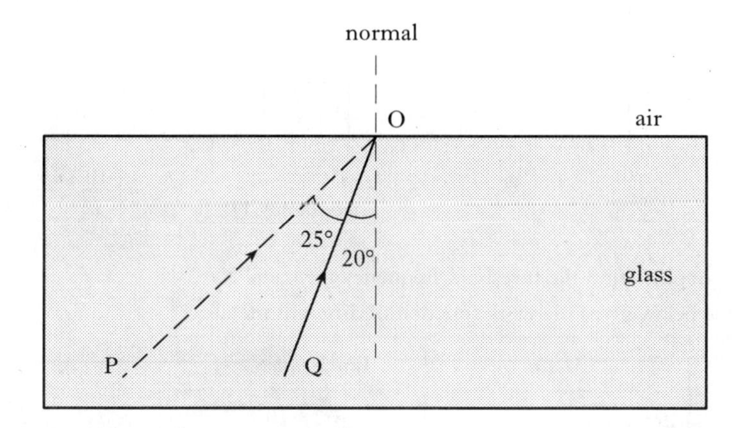

When rays PO and QO strike the glass-air boundary, **three** further rays of light are observed.

Copy and complete the diagram to show **all five** rays.

Clearly indicate which of the three rays came from P and which came from Q.

The values of all angles should be shown on the diagram. **6**

(8)

Marks

10. (*a*) The apparatus shown below is used to investigate photoelectric emission from the metal surface X when electromagnetic radiation is shone on the surface.

The frequency of the electromagnetic radiation can be varied.

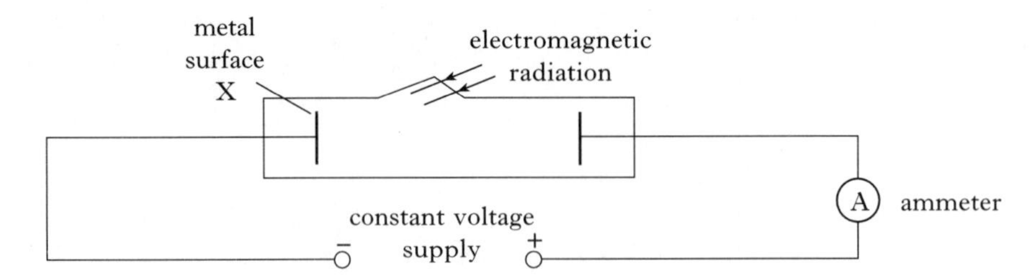

(i) When radiation of a certain frequency is shone on the metal surface X, a reading is obtained on the ammeter.

Sketch a graph to show how the current in the circuit varies with the intensity of the radiation.

(ii) Explain why there is no reading on the ammeter when the frequency of the radiation is decreased below a particular value. **3**

(*b*) The maximum kinetic energy of the photoelectrons emitted from metal X is measured for a number of different frequencies of the radiation.

The graph shows how this kinetic energy varies with frequency.

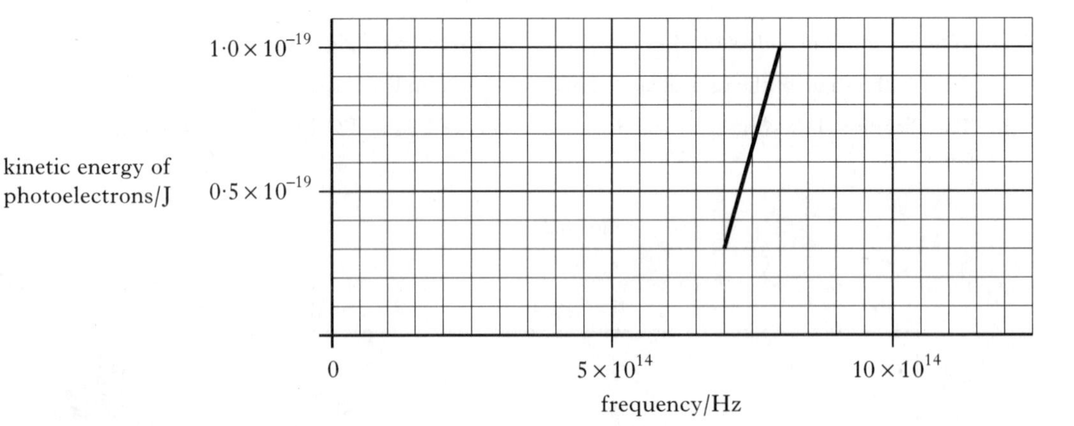

(i) Use the graph to find the threshold frequency for metal X.

(ii) The table below gives the work function of different metals.

Metal	*Work function*/J
Potassium	$3 \cdot 2 \times 10^{-19}$
Calcium	$4 \cdot 3 \times 10^{-19}$
Zinc	$6 \cdot 9 \times 10^{-19}$
Gold	$7 \cdot 8 \times 10^{-19}$

Which one of these metals was used in the investigation?

You must justify your answer using the information given in the table. **4**

(7)

Marks

11. (*a*) The diagram shows the apparatus used by Rutherford to investigate the scattering of alpha particles by a gold foil.

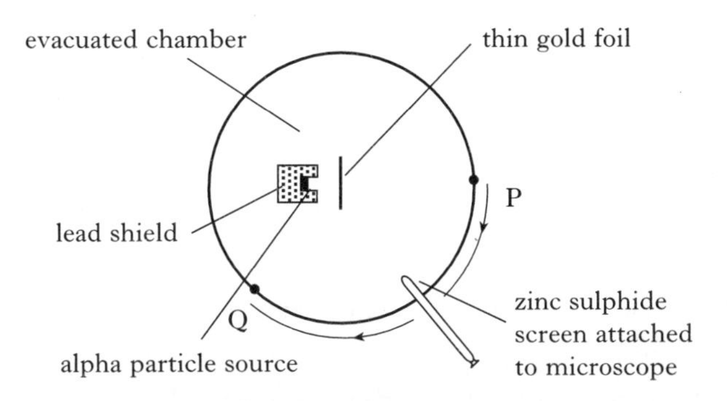

From the observations made as the microscope and screen were moved from P to Q, Rutherford deduced that an atom has a nucleus which is:

(A) positively charged;

(B) massive;

(C) much smaller than the volume of the atom.

Explain how the observations from the scattering experiment led to these three deductions. **3**

(*b*) A pupil reads in a textbook about the possible effects of a source of gamma rays and neutrons on one type of body tissue. A table in the textbook provided information relating to the radiations and absorbed doses for this body tissue. This table is shown below.

Type of radiation	*Quality factor*	*Absorbed dose*/μGy
gamma	1	200
neutrons	3	100

(i) Calculate the total dose equivalent received by the body tissue.

(ii) Calculate the thickness of lead which would have to surround the above source to reduce the absorbed dose from the gamma rays to 25 μGy.

The half-value thickness of lead for the gamma radiation is 8 mm. **5**

(8)

[*END OF QUESTION PAPER*]

SOLUTIONS — PAPER II

Fully worked solutions to these questions are given in our book entitled *Solutions to Higher Grade Physics*

SCOTTISH
CERTIFICATE OF
EDUCATION
1998

FRIDAY, 15 MAY
9.30 AM – 11.00 AM

PHYSICS
HIGHER GRADE
Paper I

Read Carefully

1 All questions should be attempted.

2 The following data should be used when required unless otherwise stated.

Speed of light in vacuum *c*	$3{\cdot}00 \times 10^{8}$ m s^{-1}	Planck's constant *h*	$6{\cdot}63 \times 10^{-34}$ J s
Charge on electron *e*	$-1{\cdot}60 \times 10^{-19}$ C	Mass of electron m_e	$9{\cdot}11 \times 10^{-31}$ kg
Acceleration due to gravity *g*	$9{\cdot}8$ m s^{-2}	Mass of proton m_p	$1{\cdot}67 \times 10^{-27}$ kg

Section A (questions 1 to 30)

3 Check that the answer sheet is for Physics Higher I (Section A).

4 Answer the questions numbered 1 to 30 on the answer sheet provided.

5 Fill in the details required on the answer sheet.

6 Rough working, if required, should be done only on this question paper, or on the first two pages of the answer book provided—**not** on the answer sheet.

7 For each of the questions 1 to 30 there is only **one** correct answer and each is worth 1 mark.

8 Instructions as to how to record your answers to questions 1–30 are given on page two.

Section B (questions 31 to 37)

9 Answer questions numbered 31 to 37 in the answer book provided.

10 Fill in the details on the front of the answer book.

11 Enter the question number clearly in the margin of the answer book beside each of your answers to questions 31 to 37.

12 Care should be taken **not** to give an unreasonable number of significant figures in the final answers to calculations.

1998

SECTION A

Answer questions 1–30 on the answer sheet.

1. Consider the following three statements made by pupils about scalars and vectors.

I Scalars have direction only.

II Vectors have both size and direction.

III Speed is a scalar and velocity is a vector.

Which statement(s) is/are true?

A I only

B I and II only

C I and III only

D II and III only

E I, II and III

2. The following is a speed-time graph of the beginning of a cyclist's journey along a straight track.

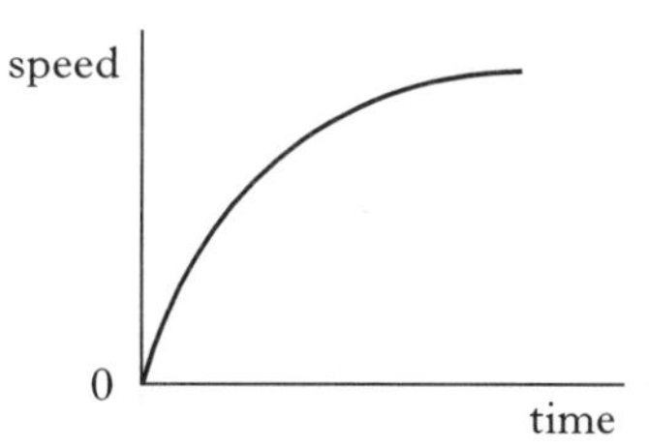

Which of the following could be the corresponding acceleration-time graph for the same period?

A

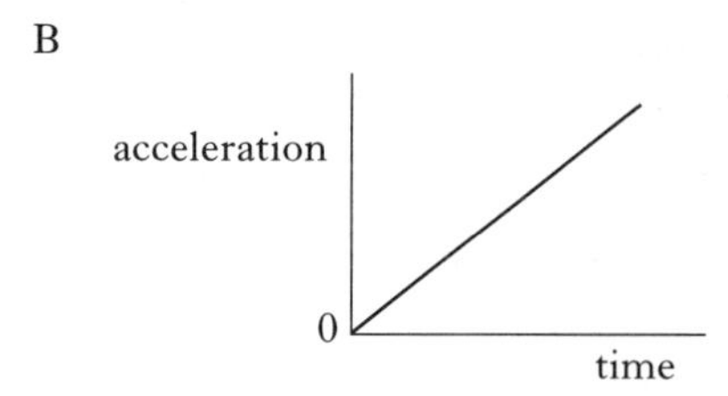

B

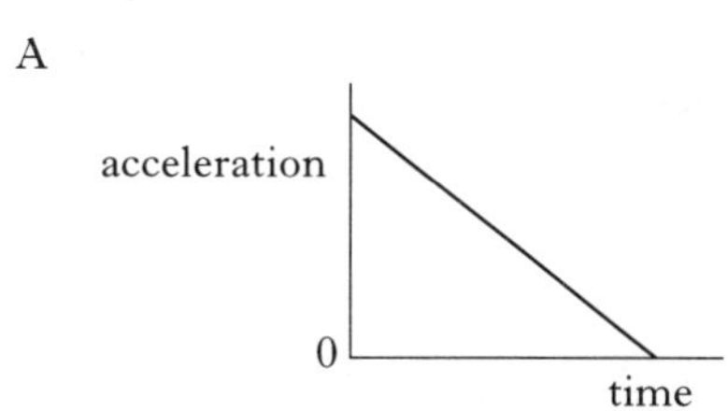

C

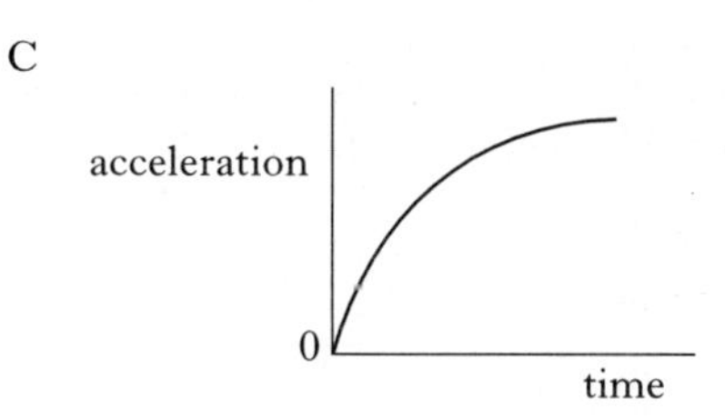

D

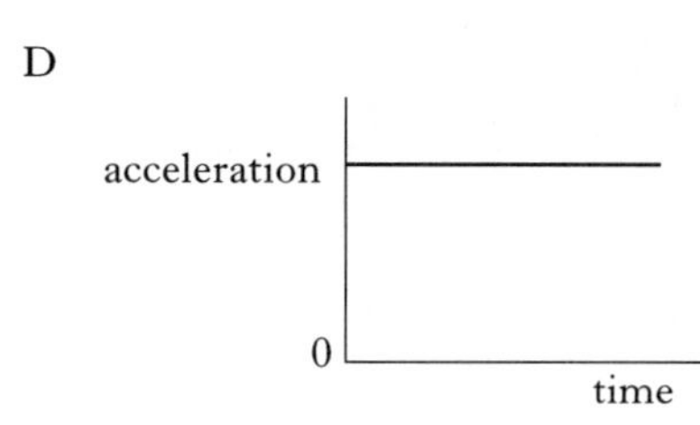

E

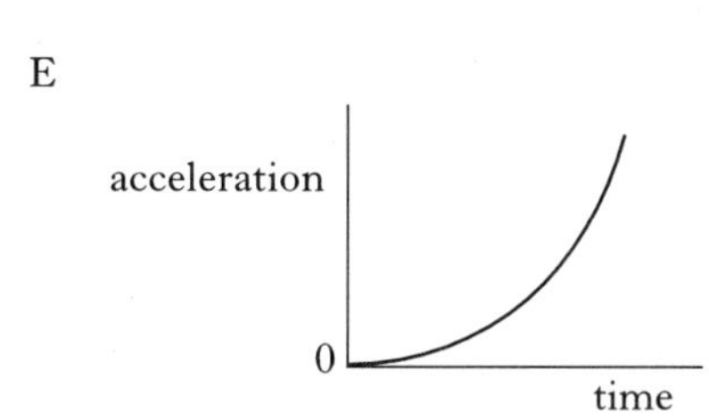

3. A cyclist is travelling along a straight, level road at 10 m s^{-1}. She applies her brakes and comes to rest after travelling a further 20 m.

The braking force is constant. What is her deceleration?

A $0{\cdot}25 \text{ m s}^{-2}$

B $0{\cdot}50 \text{ m s}^{-2}$

C $2{\cdot}0 \text{ m s}^{-2}$

D $2{\cdot}5 \text{ m s}^{-2}$

E $5{\cdot}0 \text{ m s}^{-2}$

4. A stone is thrown horizontally with a speed of 12 m s^{-1} over the edge of a vertical cliff. It hits the sea at a horizontal distance of 60 m out from the base of the cliff.

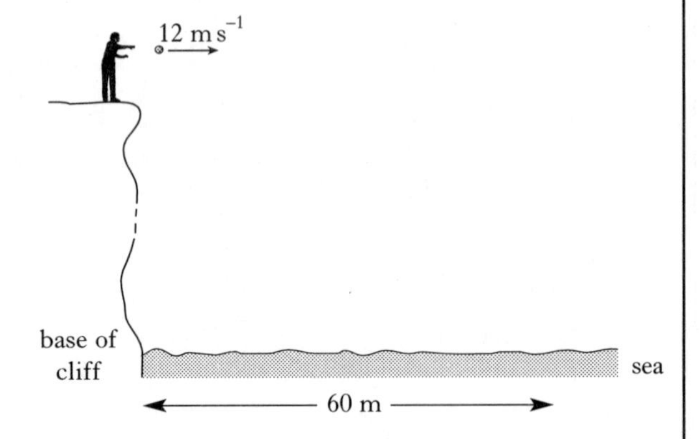

Assuming that air resistance is negligible and that the acceleration due to gravity is 10 m s^{-2}, the height from which the stone was projected above the level of the sea is

A 5 m

B 25 m

C 50 m

D 125 m

E 250 m.

5. A rocket of mass 200 kg accelerates vertically upwards from the surface of a planet at 2 m s^{-2}.

The gravitational field strength on the planet is 4 N kg^{-1}.

What is the size of the force being supplied by the rocket's engines?

A 800 N

B 1200 N

C 2000 N

D 2400 N

E 4800 N

6. Two boys are pulling a car of mass 800 kg along a level surface with a pair of ropes attached horizontally as shown below.

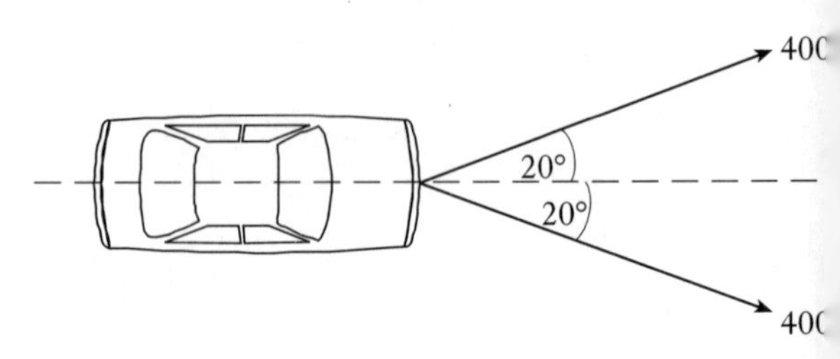

When the pull on each rope is 400 N in the directions indicated, the acceleration of the car is $0{\cdot}1 \text{ m s}^{-2}$.

What is the size of the frictional force acting on the car in the above situation?

A 194 N

B 434 N

C 533 N

D 672 N

E 832 N

7. A block of mass 1 kg slides along a frictionless surface at $10\,m\,s^{-1}$ and it collides with a stationary block of mass 10 kg. After the collision, the first block rebounds at $5\,m\,s^{-1}$ and the other one moves off at $1{\cdot}5\,m\,s^{-1}$.

before impact

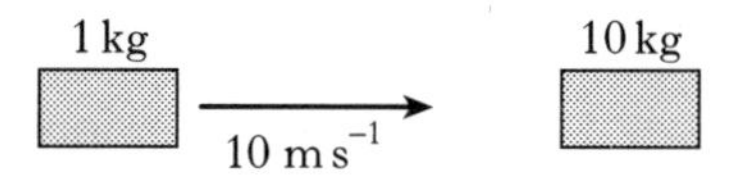

after impact

Which row in the following table is correct?

	Momentum of system	*Kinetic energy of system*	*Type of collision*
A	conserved	conserved	elastic
B	conserved	not conserved	inelastic
C	conserved	not conserved	elastic
D	not conserved	not conserved	inelastic
E	not conserved	not conserved	elastic

8. Which pair of graphs correctly shows how the pressure produced by a liquid depends on the depth and the density of the liquid?

A
pressure vs depth; pressure vs density

B
pressure vs depth; pressure vs density

C
pressure vs depth; pressure vs density

D
pressure vs depth; pressure vs density

E
pressure vs depth; pressure vs density

9. The pressure-volume graph below describes the behaviour of a constant mass of gas when it is heated.

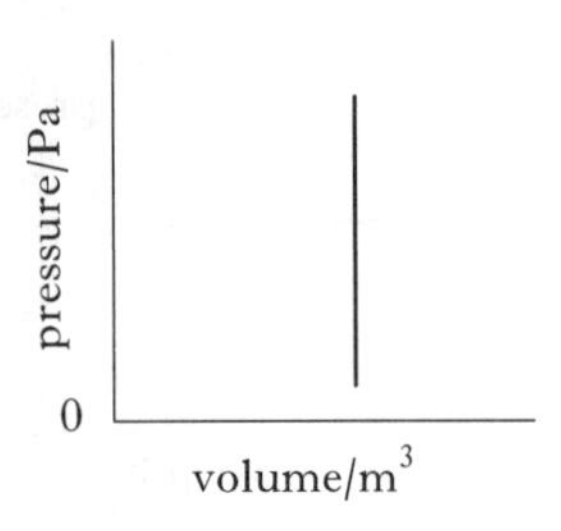

Which of the following shows the corresponding pressure-temperature graph?

A

pressure/Pa
0
temperature/K

B

pressure/Pa
0
temperature/K

C

pressure/Pa
0
temperature/K

D

pressure/Pa
0
temperature/K

E

pressure/Pa
0
temperature/K

10. A balloon of mass 10 kg accelerates vertically upwards with a constant acceleration of 1 m s^{-2}. The air resistance acting on the balloon is 100 N.

Assuming that the acceleration due to gravity is 10 m s^{-2}, which row in the following table shows the size and direction of the forces acting on the balloon?

	Weight	*Air resistance*	*Upthrust*
A	↓ 100 N	↓ 100 N	↑ 200 N
B	↓ 100 N	↓ 100 N	↑ 210 N
C	↓ 100 N	↑ 100 N	↑ 10 N
D	↓ 10 N	↓ 100 N	↑ 120 N
E	↓ 100 N	↑ 100 N	↑ 100 N

11. In the circuit below, each resistor has a resistance of 20 Ω and the battery has negligible internal resistance.

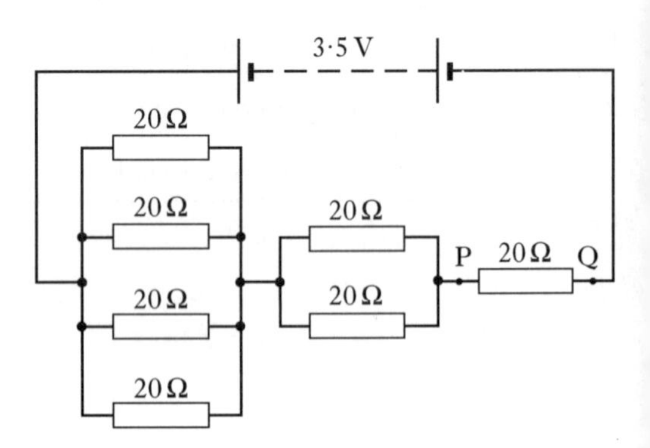

The voltage across PQ is

A 0·5 V

B 1·0 V

C 1·5 V

D 2·0 V

E 3·5 V.

12. A battery, of e.m.f. 15 V and internal resistance 5 Ω, is connected to two 10 Ω resistors as shown. Switch S is initially open.

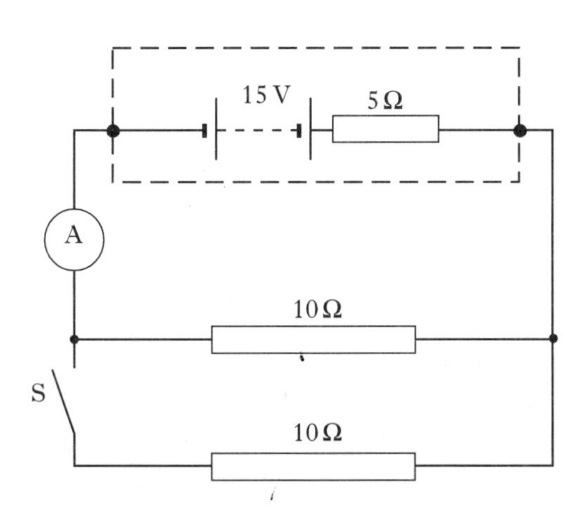

When switch S is closed, the reading on the ammeter changes

A from 1 A to 2 A

B from 1·5 A to 3 A

C from 1 A to 1·5 A

D from 1·5 A to 0·75 A

E from 1 A to 0·6 A.

13. A student sets up the following potential divider circuit which includes a light-dependent resistor (LDR).

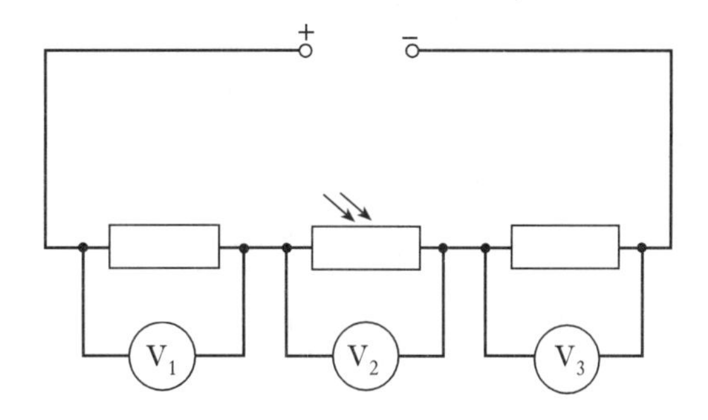

The resistance of the LDR decreases when the light intensity on it increases.

Which row in the table below correctly shows how the voltmeter readings are affected when the student switches off all the lights in the laboratory?

	Reading on voltmeter V_1	*Reading on voltmeter* V_2	*Reading on voltmeter* V_3
A	increases	increases	increases
B	decreases	decreases	decreases
C	increases	decreases	increases
D	decreases	increases	decreases
E	no change	increases	no change

14. In the Wheatstone bridge shown below, there is a small reading on the voltmeter.

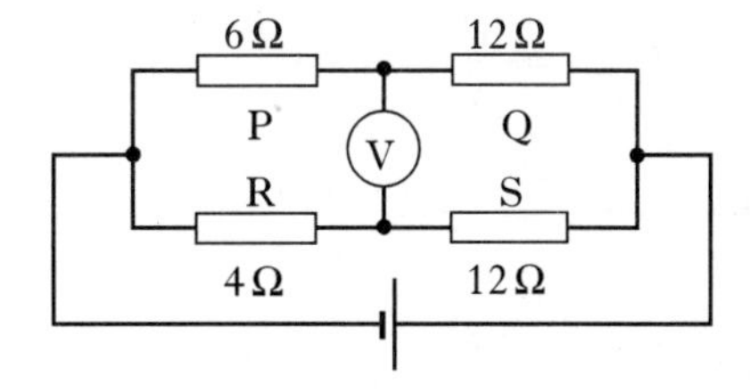

What should be done to balance the Wheatstone bridge?

A Increase the value of resistor P by 6 Ω.

B Increase the value of resistor Q by 6 Ω.

C Increase the value of resistor R by 6 Ω.

D Increase the value of resistor S by 6 Ω.

E Insert a 6 Ω resistor in series with the voltmeter.

15. The output from an electrical device produces the following trace on an oscilloscope.

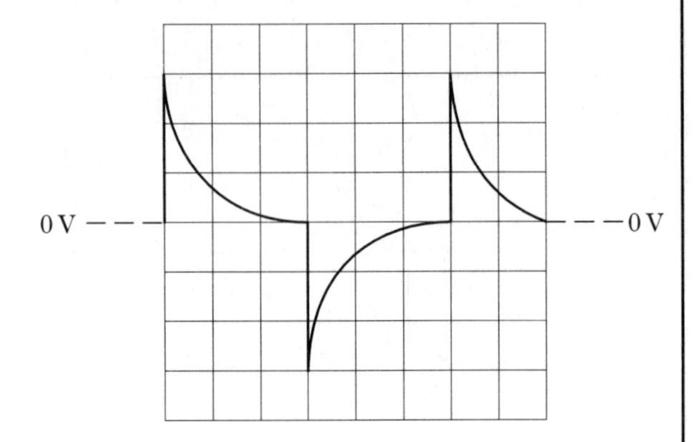

The time-base setting of the oscilloscope is 2 ms per division and the voltage-gain setting is 5 mV per division.

What is the frequency and maximum voltage of the output from this electrical device?

	Frequency of the device/Hz	*Maximum voltage output from the device*/mV
A	33	15
B	83	15
C	83	30
D	166	30
E	166	15

16. What is the relationship between the r.m.s. and peak values for an alternating current?

A $I_{r.m.s.} = \frac{I_p}{\sqrt{2}}$

B $I_{r.m.s.} = \sqrt{2}\, I_p$

C $I_{r.m.s.} = 2\, I_p^2$

D $I_{r.m.s.} = \frac{\sqrt{I_p}}{2}$

E $I_{r.m.s.} = \frac{I_p^2}{2}$

17. A 25 μF capacitor is charged until the potential difference across it is 500 V. The charge stored in the capacitor is

A $5{\cdot}00 \times 10^{-8}$ C

B $2{\cdot}00 \times 10^{-5}$ C

C $1{\cdot}25 \times 10^{-2}$ C

D $1{\cdot}25 \times 10^{4}$ C

E $2{\cdot}00 \times 10^{7}$ C.

18. The graph shows how the charge stored on a capacitor varies as the p.d. applied across it is increased.

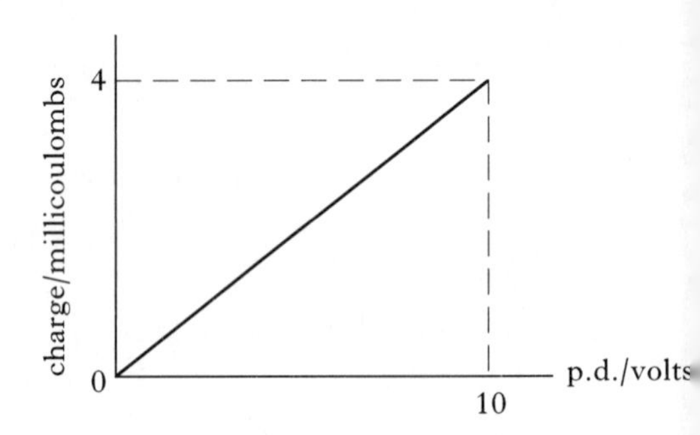

What is the energy stored in the capacitor when the p.d. across it is 10 V?

A 0·4 mJ

B 2·5 mJ

C 10 mJ

D 20 mJ

E 40 mJ

19. In the circuits shown below, P and Q are identical lamps and the a.c. supplies have the same r.m.s. voltage output. The lamps glow with equal brightness.

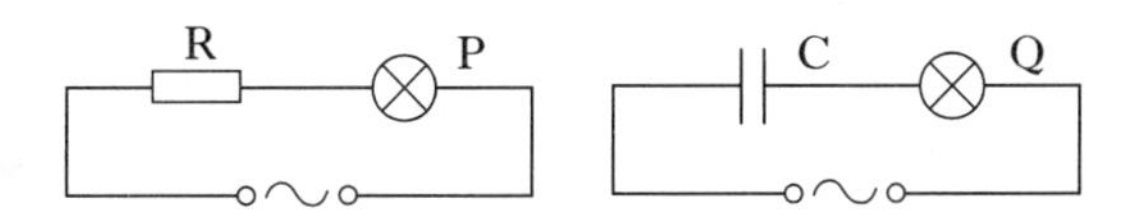

The frequency of each supply voltage is increased without altering the value of the r.m.s. voltage output.

Which row in the following table correctly describes how the brightness of each lamp is affected?

	Lamp P	*Lamp Q*
A	brighter	unchanged
B	unchanged	brighter
C	dimmer	unchanged
D	unchanged	dimmer
E	dimmer	brighter

20. The amplifier shown below is operating in the differential mode.

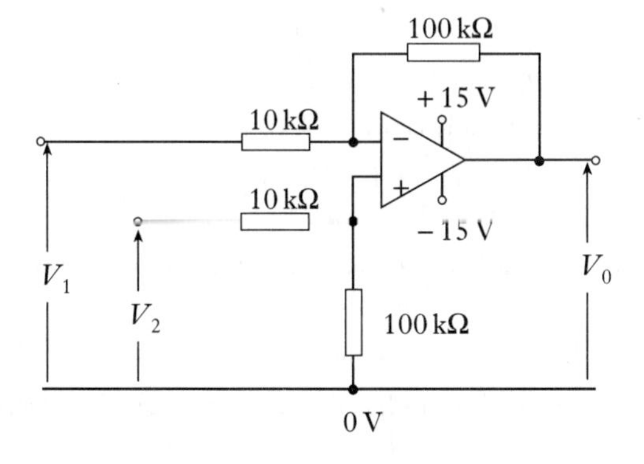

When $V_0 = 0{\cdot}60$ V and $V_1 = 2{\cdot}70$ V, what is the value of V_2?

A 2·10 V

B 2·16 V

C 2·76 V

D 3·30 V

E 3·36 V

21. The diagram below shows a ray of light from a laser passing from air into glass and then into water.

The refractive index of glass is greater than that of water.

Which is the correct path for the light?

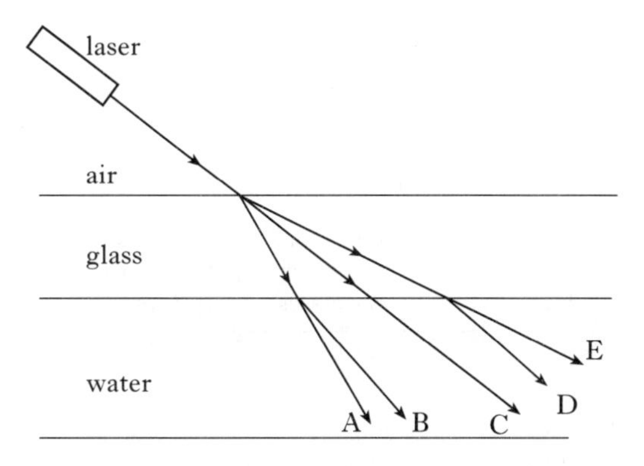

22. A ray of light travelling through glass approaches air, as shown below.

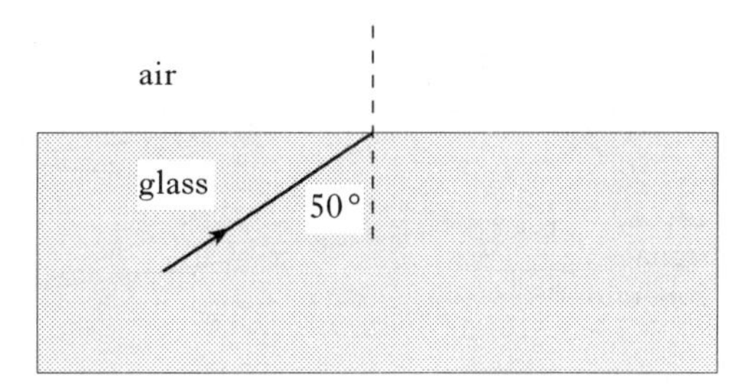

The refractive index of the glass is 1·5.

Which of the following paths will the ray follow?

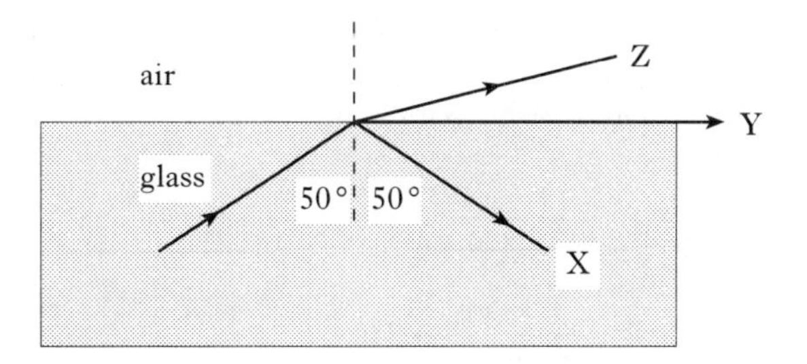

A X only

B Y only

C Z only

D X and Z only

E Y and Z only

23. Light travels from air into glass.

Which row in the following table correctly describes what happens to the speed, frequency and wavelength of the light?

	Speed	*Frequency*	*Wavelength*
A	increases	decreases	stays constant
B	decreases	stays constant	decreases
C	stays constant	decreases	decreases
D	increases	stays constant	increases
E	decreases	decreases	stays constant

24. Two identical loudspeakers, L_1 and L_2, are operated at the same frequency and in phase with each other by connecting them in parallel across the output of a signal generator, as shown below. A sound interference pattern is produced.

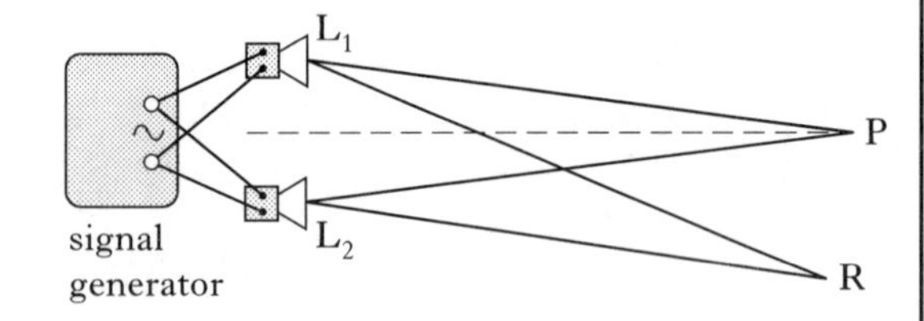

At position P, which is the same distance from both loudspeakers, a microphone registers a maximum intensity of sound.

The next maximum is registered at position R, where $L_1R = 4{\cdot}6$ m and $L_2R = 4{\cdot}3$ m.

If the speed of sound is 330 m s^{-1}, then the frequency of the sound emitted by the loudspeakers is given by

A $\dfrac{(4{\cdot}6 - 4{\cdot}3)}{330}$ Hz

B $\dfrac{330}{(4{\cdot}6 + 4{\cdot}3)}$ Hz

C $\dfrac{330}{(4{\cdot}6 - 4{\cdot}3)}$ Hz

D $330 \times (4{\cdot}6 - 4{\cdot}3)$ Hz

E $330 \times (4{\cdot}6 + 4{\cdot}3)$ Hz

25. The intensity of radiation emitted from a point source of light varies

A directly as the distance from the source

B directly as the square of the distance from the source

C directly as the square root of the distance from the source

D inversely as the distance from the source

E inversely as the square of the distance from the source.

26. Certain materials can be "doped" to make a semiconductor called an n-type material.

In an n-type material,

A the majority charge carriers are electrons

B the majority charge carriers are neutrons

C the majority charge carriers are protons

D there are more electrons than protons

E there are more electrons than neutrons.

27. The symbols for two isotopes of carbon, carbon 14 and carbon 12, are as follows.

$^{14}_{6}\mathrm{C}$ $\qquad$ $^{12}_{6}\mathrm{C}$

Which of the following statements is true?

Carbon 14 and carbon 12 are said to be isotopes of carbon because

A carbon 14 has the same mass number as carbon 12

B carbon 14 has a different atomic number from carbon 12

C carbon 14 is radioactive

D carbon 14 has the same number of neutrons as carbon 12

E carbon 14 and carbon 12 have different mass numbers but the same atomic number.

28. Two different types of tinted glass, X and Y, are used to make filters for sunglasses. A sample of each glass is placed in turn between a source and a detector, as shown in the following diagram. Both samples are identical in size and shape.

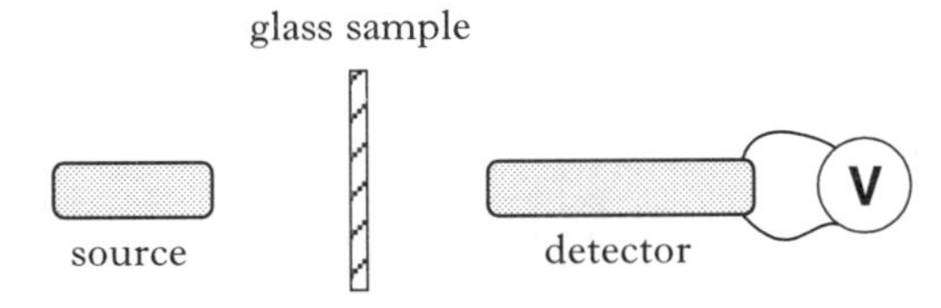

The source emits electromagnetic radiation with a wide range of wavelengths, all of the same intensity as shown below.

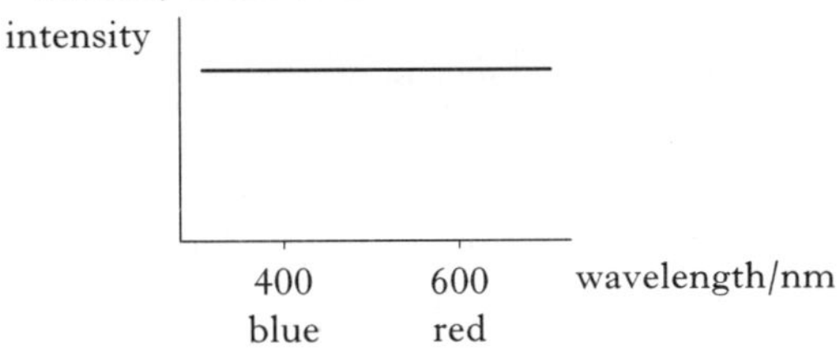

The following graphs show the intensities measured by the detector after the electromagnetic radiation passed through each of the glass samples.

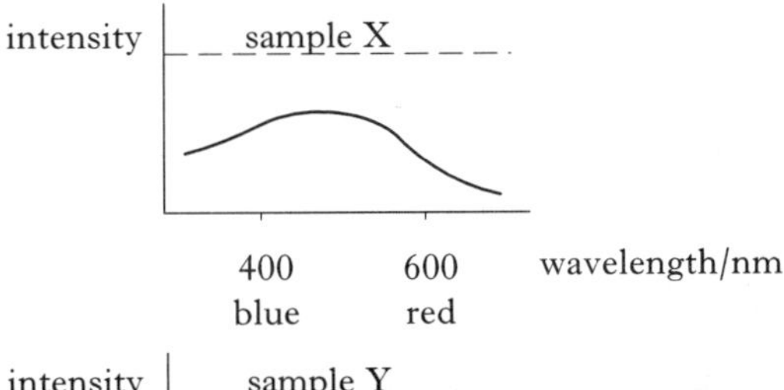

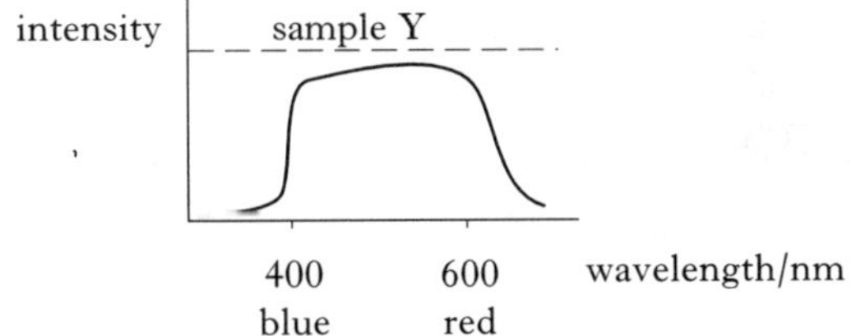

Which of the following statement(s) is/are correct?

I Sample X is better at absorbing red light than sample Y.

II Sample Y is better at protecting the eye from ultraviolet radiation.

III The view would appear darker when seen through sample Y than when seen through sample X.

A I only

B I and II only

C I and III only

D II and III only

E I, II and III

29. There are a number of equations involving the following quantities.

A, the activity of a radioactive source

D, the absorbed dose

H, the dose equivalent

$\dot{H}$, the dose equivalent rate

Q, the quality factor

N, the number of nuclei decaying

t, the time

Which row of the following table states **three** of these equations correctly?

A	$A = \frac{N}{t}$	$H = DQ$	$\dot{H} = Ht$
B	$A = Nt$	$H = DQ$	$\dot{H} = \frac{H}{t}$
C	$A = \frac{N}{t}$	$H = \frac{D}{Q}$	$\dot{H} = Ht$
D	$A = Nt$	$H = \frac{D}{Q}$	$\dot{H} = \frac{H}{t}$
E	$A = \frac{N}{t}$	$H = DQ$	$\dot{H} = \frac{H}{t}$

30. The three statements below refer to the fission process.

I Fission may be spontaneous.

II Fission can be produced when neutrons bombard a nucleus, which has a large mass number.

III When fission occurs, a nucleus with a large mass number may split into nuclei with smaller mass numbers, along with several neutrons.

Which statement(s) is/are true?

A III only

B I and II only

C I and III only

D II and III only

E I, II and III

SECTION B

Write your answers to questions 31 to 37 in the answer book.

Marks

31. A spectator at A walks to C, the opposite corner of a playing field, by walking from A to B and then from B to C as shown in the diagram below.

The distance from A to B is 50 m. The distance from B to C is 150 m.

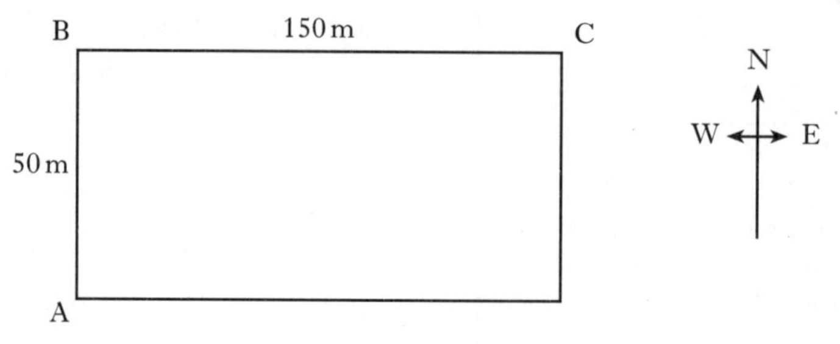

By scale drawing or otherwise, find the resultant displacement. Magnitude and direction are required. **2**

32. A football of mass 0·42 kg is thrown at a stationary student of mass 50·0 kg who is wearing roller blades, as shown in the diagram below. When the student catches the moving ball she moves to the right.

The instantaneous speed immediately after she catches the ball is 0·10 $m\,s^{-1}$.

Calculate the speed of the ball just before it is caught. **2**

33. Calculate the size of the current in the ammeter in the circuit below. The battery has negligible internal resistance.

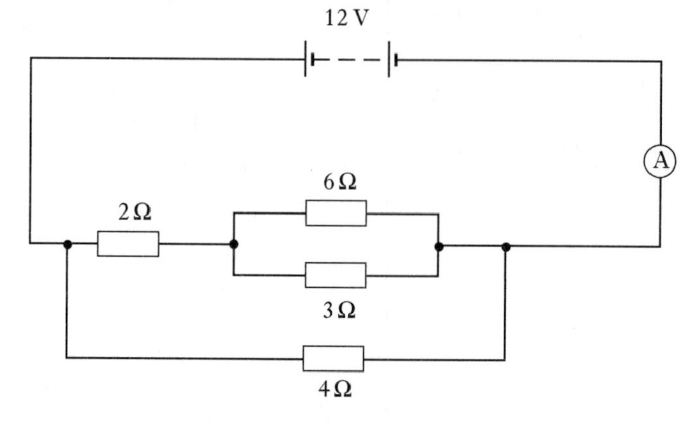

3

Marks

34. The Wheatstone bridge shown below is balanced.

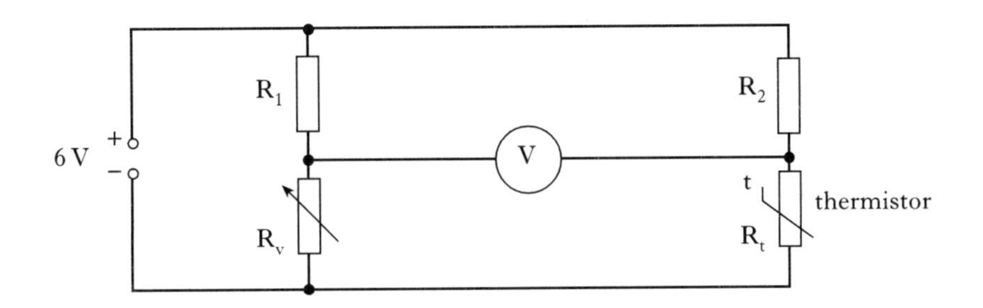

(*a*) R_1 has a resistance of 3·3 kΩ, R_2 has a resistance of 2·2 kΩ and the variable resistor R_v is set at 225 Ω. Calculate the resistance of the thermistor R_t.

(*b*) The graph below shows what happens to the reading on the voltmeter as the temperature of thermistor R_t is changed.

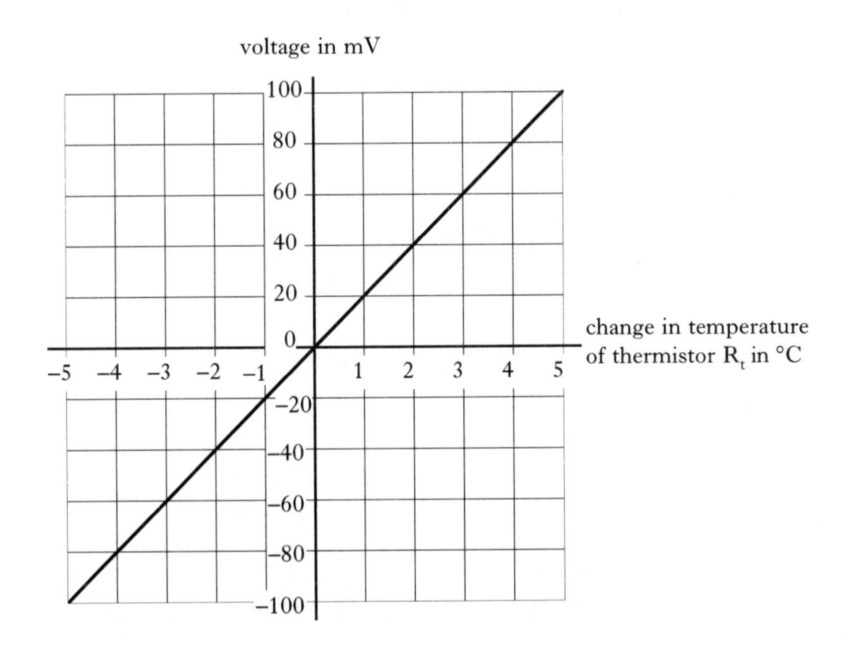

The bridge was initially balanced at 20 °C.

The temperature of R_t is increased until the reading on the voltmeter is 80 mV. What is the new temperature of the thermistor R_t? **3**

35. The minimum energy required to cause an electron to be emitted from a clean zinc surface is $6{\cdot}9 \times 10^{-19}$ J.

(*a*) Calculate the maximum wavelength of electromagnetic radiation which will cause an electron to be emitted from the clean zinc surface.

(*b*) What would be the effect of irradiating a clean zinc surface with radiation of wavelength $4{\cdot}0 \times 10^{-7}$ m? You must justify your answer. **4**

36. The diagram shows a simplified view of a gas laser. *Marks*

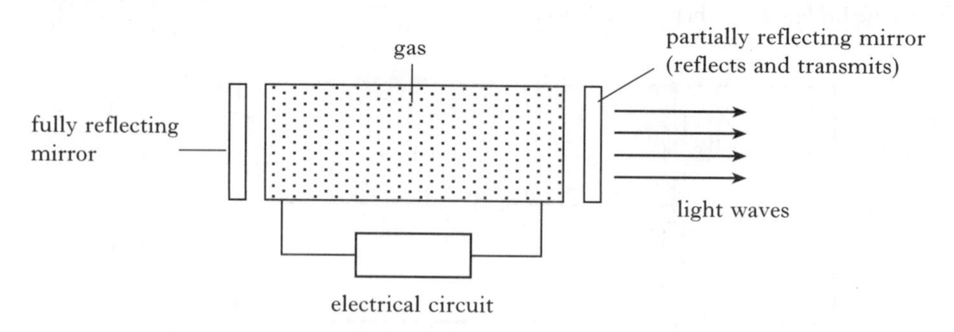

Laser stands for Light Amplification by Stimulated Emission of Radiation.

(*a*) Explain what is meant by *stimulated emission of radiation.*

(*b*) State **two** ways in which the incident radiation and the radiation it stimulates are similar. **3**

37. A grating with 300 lines/mm is used with a spectrometer and a source of monochromatic light to view an interference pattern as shown below.

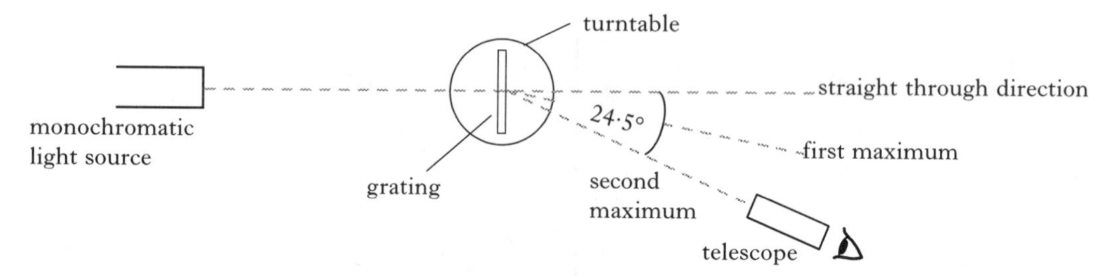

The second maximum of interference is observed when the telescope is at an angle of 24·5°. Calculate the wavelength of the light. **3**

[*END OF QUESTION PAPER*]

HIGHER PHYSICS ANSWERS — Paper I

Section A

1. D	**2.** A	**3.** D	**4.** D	**5.** B	**6.** D	**7.** B	**8.** E	**9.** C	**10.**
11. D	**12.** C	**13.** D	**14.** B	**15.** B	**16.** A	**17.** C	**18.** D	**19.** B	**20.**
21. B	**22.** A	**23.** B	**24.** C	**25.** E	**26.** A	**27.** E	**28.** B	**29.** E	**30.**

31. 158 m at 72° E of N.

32. Show working to give speed of ball = 12 m s^{-1}

33. Show working to give 6 A.

34. *(a)* Show working to give 150 ohms. *(b)* Show working to give 24 °C.

35. *(a)* Show working to give 3×10^{-7} m.

(b) No photoelectric emissions since $E = h \times f$ is not big enough.

36. *(a)* Electrons in higher level are brought to lower energy level by a photon of energy equal to the difference in energy levels. This produces identical photons.

(b) In phase; same direction.

37. Show working to give $6{\cdot}91 \times 10^{-7}$ m.

SCOTTISH CERTIFICATE OF EDUCATION 1998

FRIDAY, 15 MAY
1.00 PM – 3.30 PM

PHYSICS HIGHER GRADE Paper II

Read carefully

1 All questions should be attempted.

2 Enter the question number clearly in the margin beside each question.

3 Any necessary data will be found in the Data Sheet on page two.

4 Care should be taken not to give an unreasonable number of significant figures in the final answers to calculations.

5 Square-ruled paper (if used) should be placed inside the front cover of the answer book for return to the Scottish Qualifications Authority.

Marks

1. A trolley of mass 2·0 kg is catapulted up a slope. The slope is at an angle of 20° to the horizontal as shown in the diagram below. The speed of the trolley when it loses contact with the catapult is 3·0 $m\,s^{-1}$.

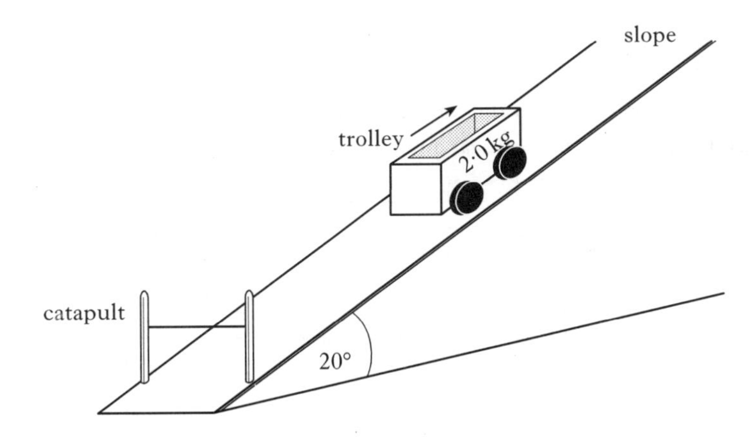

The size of the force of friction acting on the trolley as it moves up the slope is 1·3 N.

(*a*) (i) Calculate the component of the weight of the trolley acting parallel to the slope.

(ii) Draw a diagram to show the forces acting on the trolley as it moves **up the slope** and is no longer in contact with the catapult.

Show only forces or components of forces acting parallel to the slope. Name the forces.

(iii) Show that, as the trolley moves up the slope, it has a deceleration of magnitude 4·0 $m\,s^{-2}$.

(iv) Calculate the time taken for the trolley to reach its furthest point up the slope.

(v) Calculate the maximum distance the trolley travels along the slope. **9**

The trolley now moves back down the slope.

(*b*) (i) Draw a diagram to show the forces acting on the trolley as it moves **down the slope**.

Show only forces or components of forces acting parallel to the slope. Name the forces.

(ii) The magnitude of the deceleration of the trolley is 4·0 $m\,s^{-2}$ as it moves up the slope. Explain why the magnitude of the acceleration is not 4·0 $m\,s^{-2}$ when the trolley moves down the slope. **2**

(11)

Marks

2. A student performs an experiment to study the motion of the school lift as it moves upwards.

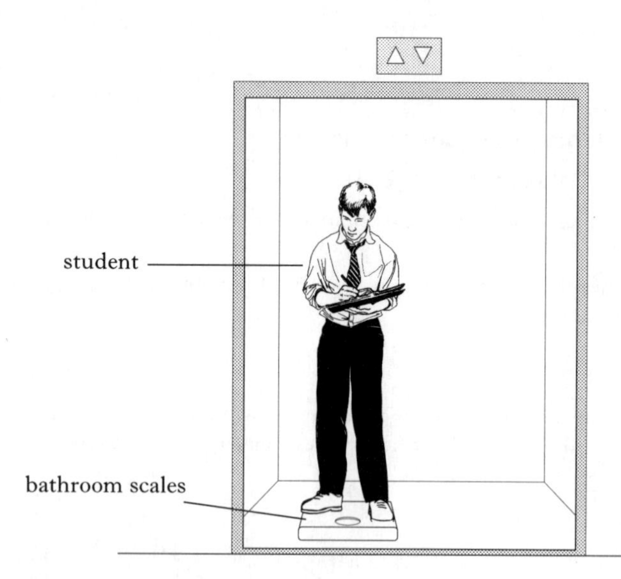

The student stands on bathroom scales during the lift's journey upwards.
The student records the reading on the scales at different parts of the lift's journey as follows.

Part of journey	*Reading on scales*
At the start (while the lift is accelerating)	678 N
In the middle (while the lift is moving at a steady speed)	588 N
At the end (while the lift is decelerating)	498 N

(*a*) Show that the mass of the student is 60 kg. **2**

(*b*) Calculate the initial acceleration of the lift. **2**

(*c*) Calculate the deceleration of the lift. **1**

(*d*) During the journey, the lift accelerates for 1·0 s, moves at a steady speed for 3·0 s and decelerates for a further 1·0 s before coming to rest.
Sketch the acceleration-time graph for this journey. **2**

(7)

Marks

3. The apparatus in the diagram is being used to investigate the average force exerted by a golf club on a ball.

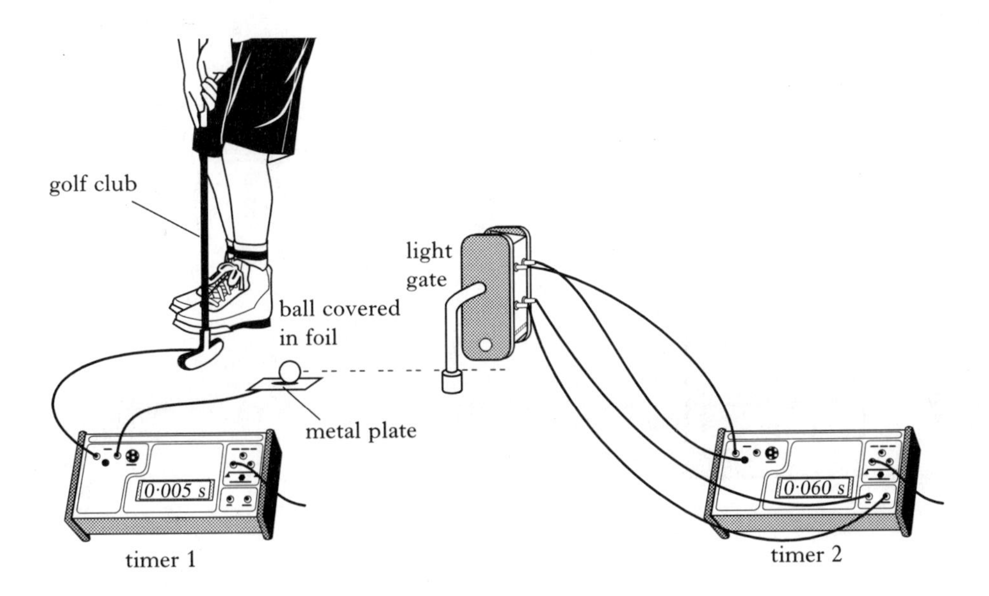

The club hits the stationary ball. Timer 1 records the time of contact between the club and the ball. Timer 2 records the time taken for the ball to pass through the light gate beam.

The mass of the ball is 45·00 ±0·01 g.

The time of the contact between club and ball is 0·005 ±0·001 s.

The time for the ball to pass through the light gate beam is 0·060 ±0·001 s.

The diameter of the ball is 24 ±1 mm.

(*a*) (i) Calculate the speed of the ball as it passes through the light gate.

(ii) Calculate the average force exerted on the ball by the golf club. **3**

(*b*) (i) Show by calculation which measurement contributes the largest percentage error in the final value of the average force on the ball.

(ii) Express your numerical answer to (*a*)(ii) in the form

final value ±absolute error. **3**

(6)

Marks

4. The rigid container of a garden sprayer has a total volume of 8·0 litres ($8 \times 10^{-3}\ m^3$).
A gardener pours 5·0 litres ($5 \times 10^{-3}\ m^3$) of water into the container. The pressure of the air inside the container is $1{\cdot}01 \times 10^5$ Pa.

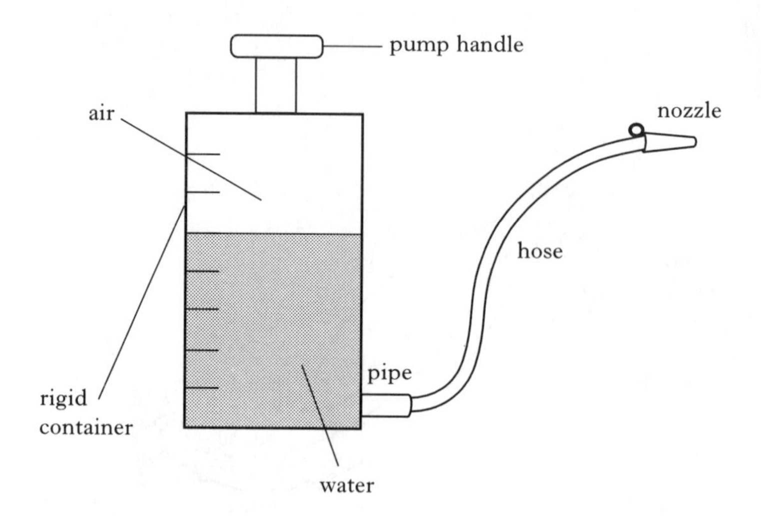

(*a*) Calculate the mass of air in the sprayer. Use information from the data sheet. **3**

(*b*) The gardener now pumps air into the container until the pressure of the air inside it becomes $3{\cdot}0 \times 10^5$ Pa.

(i) The area of the water surface in contact with the compressed air is $7{\cdot}0 \times 10^{-3}\ m^2$.
Calculate the force which the compressed air exerts on the water.

(ii) Water is now released through the nozzle. Calculate the final pressure of the air inside the sprayer when the volume of water falls from 5·0 litres ($5 \times 10^{-3}\ m^3$) to 2·0 litres ($2 \times 10^{-3}\ m^3$).
Assume the temperature of the compressed air remains constant. **4**

(7)

Marks

5. (*a*) A cell of e.m.f. 1·5 V and internal resistance 0·75 Ω is connected as shown in the following circuit.

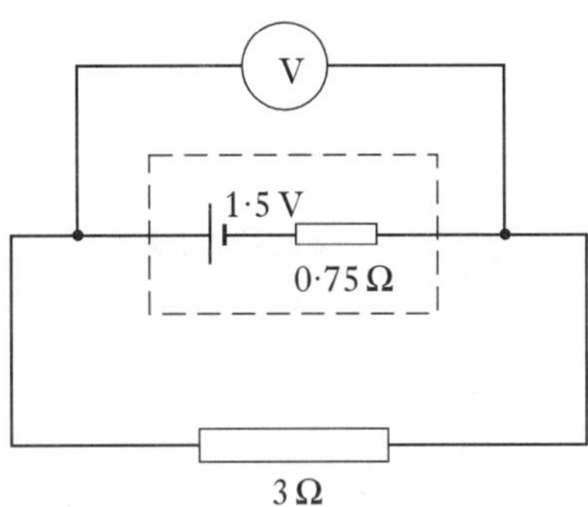

(i) Calculate the value of the reading on the voltmeter.

(ii) What is the value of the "lost volts" in this circuit? **5**

(*b*) A battery of e.m.f. 6 V and internal resistance, *r*, is connected to a variable resistor R as shown in the following circuit diagram.

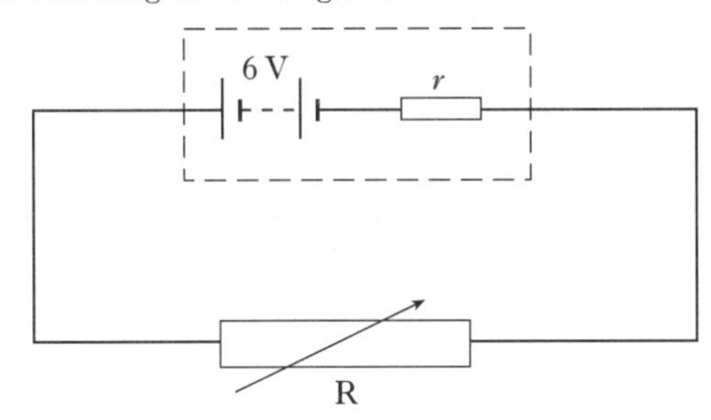

The graph below shows how the "lost volts" of this battery changes as the resistance of R increases.

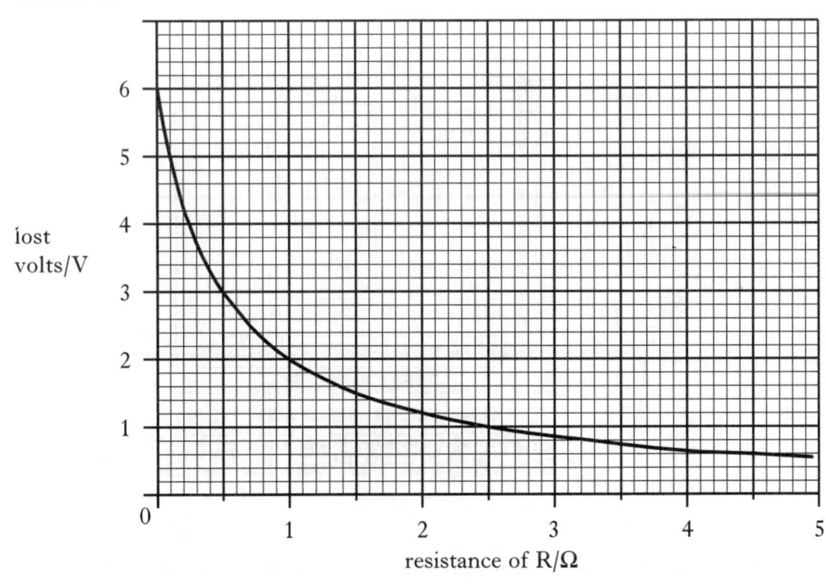

(i) Use information from the graph to calculate the p.d. across the terminals of the battery (t.p.d.) when the resistance of R is 1 Ω.

(ii) Calculate the internal resistance, *r*, of the battery. **4**

(9)

Marks

6. (*a*) A capacitor has a value of 5 μF. Explain in terms of electric charge what this means. **1**

(*b*) The 5 μF capacitor shown in the circuit below is initially uncharged. The circuit is connected to a computer and switch S is closed. The monitor of the computer displays a graph of current against time as the capacitor charges.

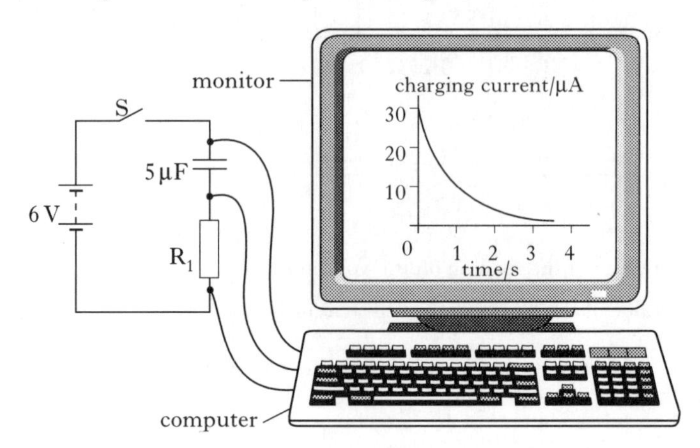

The battery has negligible internal resistance.

(i) Calculate the resistance of R_1.

(ii) The resistor R_1 is replaced by another resistor R_2. The resistance of R_2 is half that of R_1.

The capacitor is discharged and the experiment repeated.

Sketch the graph of charging current against time when R_2 is used. Include values on the axes. **3**

(*c*) In the following circuit a variable resistor R is used to keep the current constant as a different capacitor charges. The graphs on the monitor show how the charging current and p.d. across the capacitor vary with time after switch S is closed.

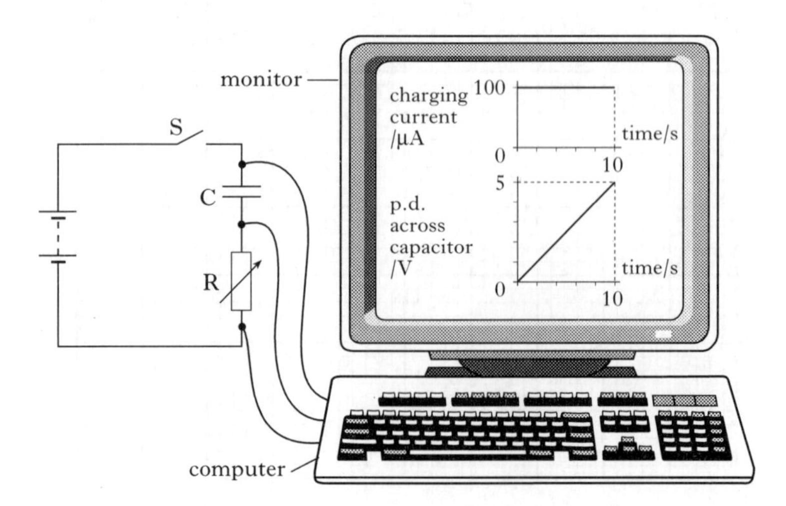

(i) What adjustment must be made to the variable resistor R so that a constant charging current is produced?

(ii) Show by calculation that 10 seconds after switch S is closed, the charge on the capacitor is 1mC.

(iii) Calculate the capacitance of C. **4**

(8)

Marks

7. The diagram below shows a cathode ray tube used in an oscilloscope.

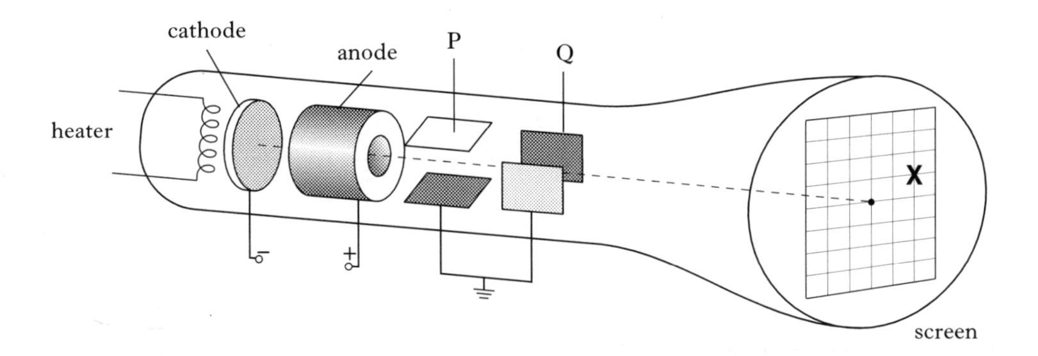

The electrons which are emitted from the cathode start from rest and reach the anode with a speed of $4{\cdot}2 \times 10^7\ \mathrm{m\,s^{-1}}$.

(*a*) (i) Calculate the kinetic energy in joules of each electron just before it reaches the anode.

(ii) Calculate the p.d. between the anode and the cathode. **4**

(*b*) Describe how the spot at the centre of the screen produced by the electrons can be moved to position **X**.

Your answer must make reference to the relative sizes and polarity (signs) of the voltages applied to plates P and Q. **2**

(6)

Marks

8. An op-amp is connected in an amplifier circuit as shown below.

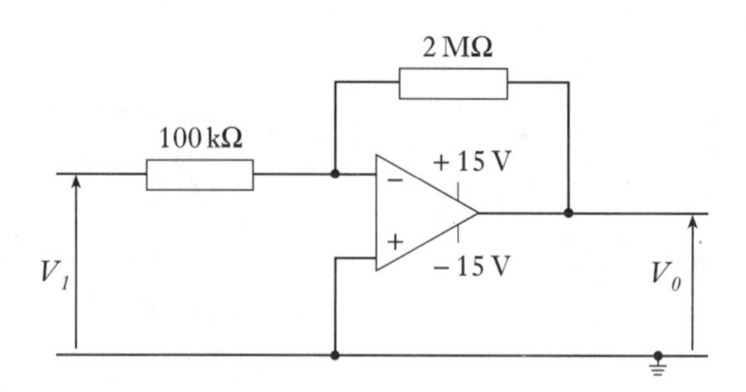

(*a*) (i) State the mode in which the op-amp is working.

(ii) Calculate the gain of this amplifier circuit.

(iii) The following graph shows how the input voltage V_1 varies with time.

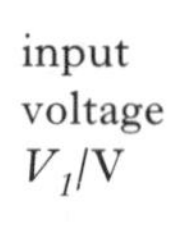

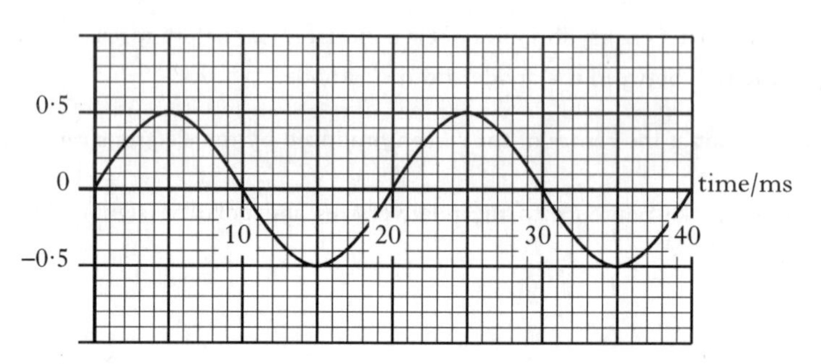

Sketch a graph to show how the output voltage V_0 varies with time. **5**

(*b*) The amplifier circuit above is modified to give the following output voltage.

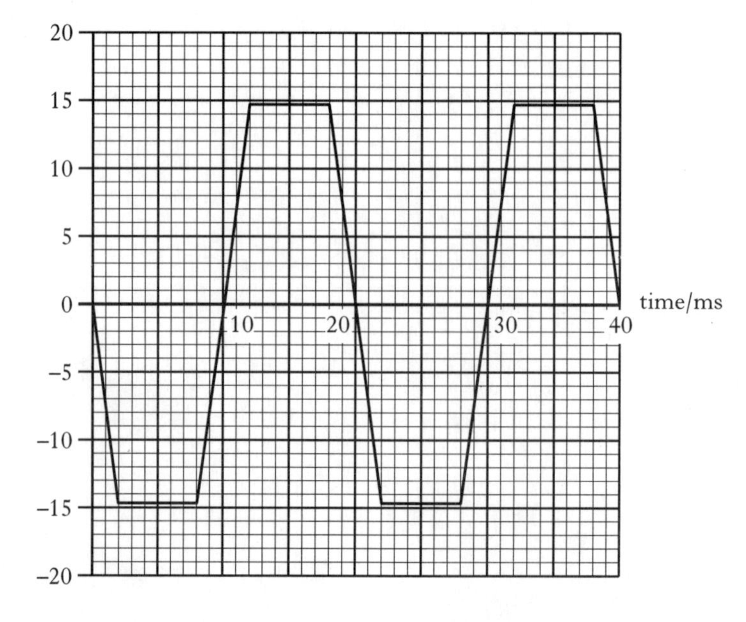

(i) Explain the shape of the output voltage graph between 2 ms and 8 ms.

(ii) Describe **two** alterations which could be made to the circuit above to give this output voltage. **3**

(8)

Marks

9. The line emission spectrum of hydrogen has four lines in the visible spectrum as shown in the following diagram.

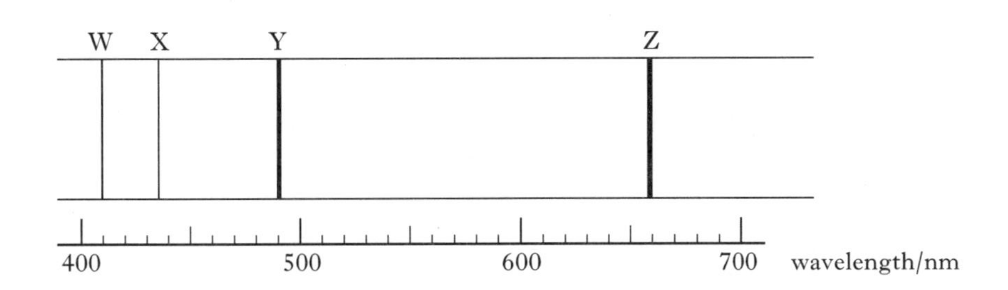

These four lines are caused by electron transitions in a hydrogen atom from high energy levels to a low energy level E_2 as shown below.

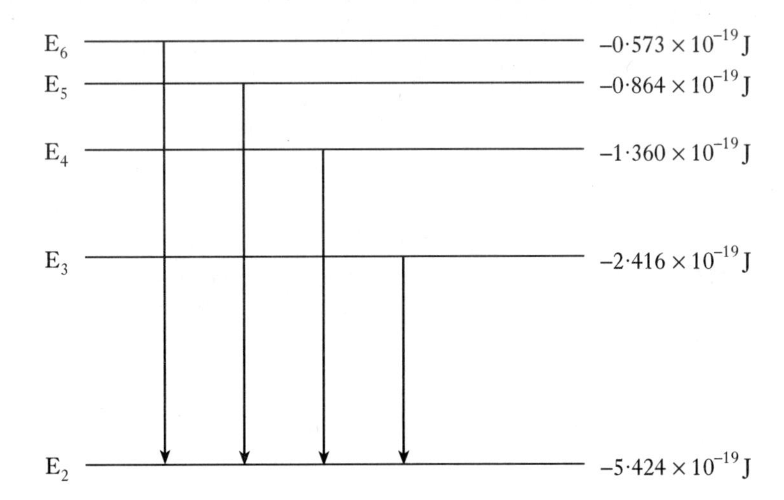

E_1 $-21{\cdot}760 \times 10^{-19}$ J

(*a*) From the information above, state which spectral line W, X, Y or Z is produced by an electron transition from E_3 to E_2. **1**

(*b*) Explain why lines Y and Z in the line emission spectrum are brighter than the other two lines. **1**

(*c*) Infrared radiation of frequency $7{\cdot}48 \times 10^{13}$ Hz is emitted from a hydrogen atom.

(i) Calculate the energy of one photon of this radiation.

(ii) Show by calculation which electron transition produces this radiation. **4**

(6)

Marks

10. (*a*) The following diagram shows a ray of monochromatic light passing from air into a block of borate glass.

The diagram is drawn to scale.

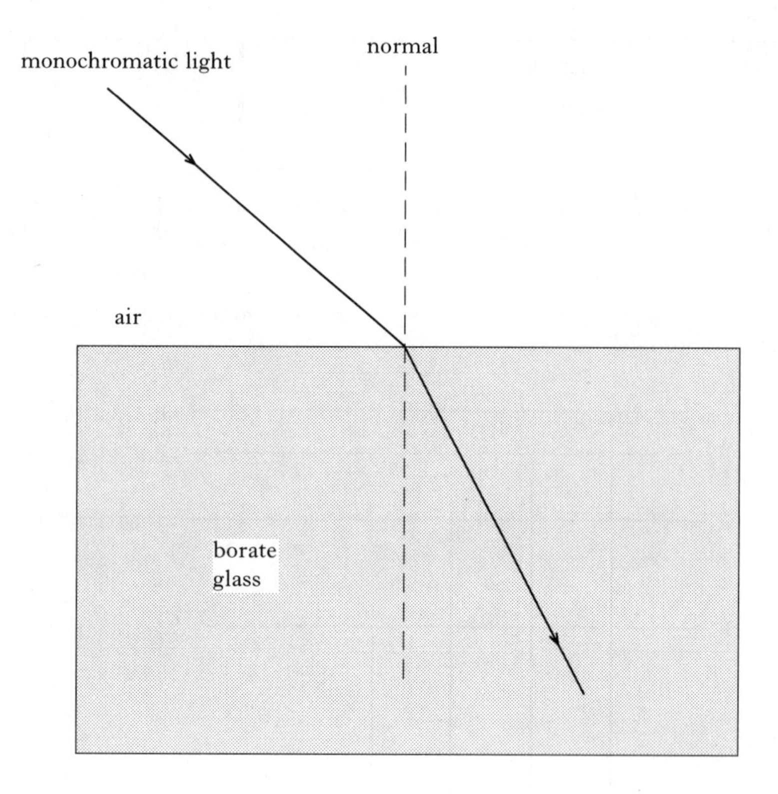

(i) Use measurements taken from the above diagram to calculate the refractive index of borate glass for this light. You will need to use a protractor.

(ii) Calculate the value of the critical angle for this light in the borate glass. **4**

Marks

(*b*) The following graph shows how refractive index depends on the type of material and the wavelength in air of the light used.

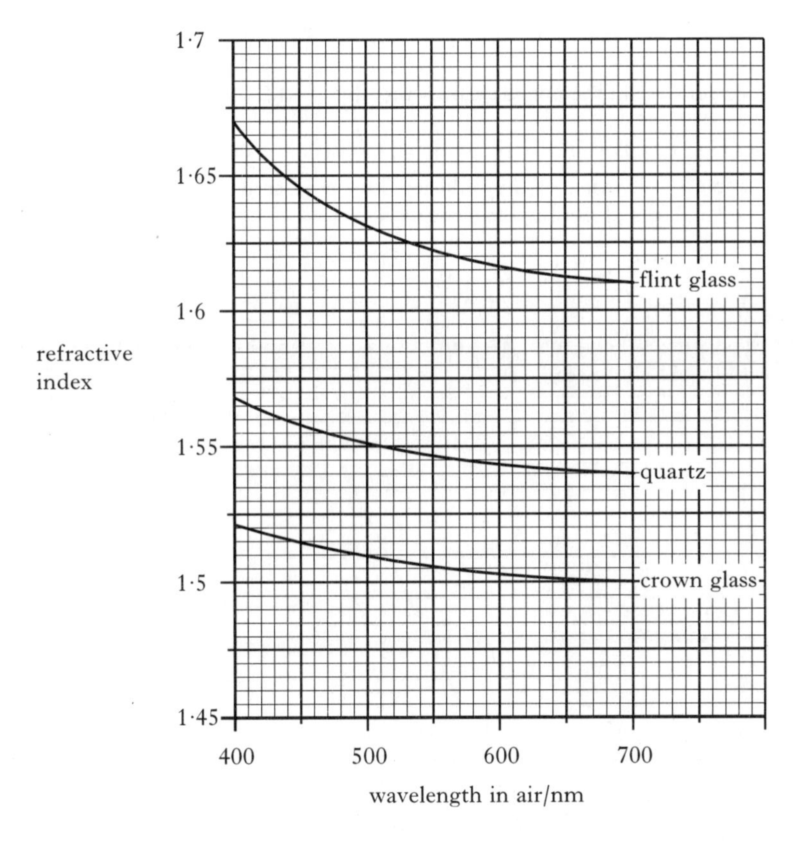

A ray of light of wavelength 510 nm in air passes into a block of quartz.

(i) Calculate the wavelength of this light in the quartz.

(ii) Explain what happens to the value of the critical angle in quartz as the wavelength of visible light increases.

(iii) A ray of white light enters a triangular prism made of crown glass, producing a visible spectrum on a screen, as shown below.

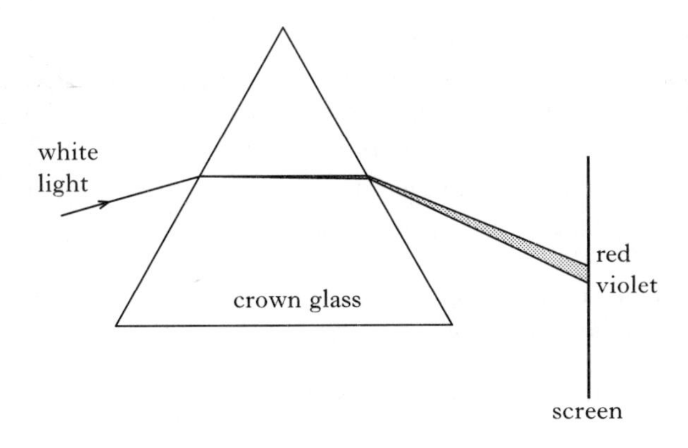

The crown glass prism is now replaced by a similar prism made from flint glass.

Describe how the visible spectrum on the screen will be different from before. **5**

(9)

11. (*a*) The first three stages in a radioactive decay series are shown below. *Marks*

$$^{238}_{92}U \longrightarrow\ ^{234}_{90}Th \longrightarrow\ ^{234}_{91}Pa \longrightarrow\ ^{234}_{92}U$$

(i) What particle is emitted when Thorium (Th) decays to Palladium (Pa)?

(ii) How many neutrons are in the nuclide represented by $^{238}_{92}U$?

(iii) In the next stage of the above decay series, an alpha particle is emitted.

Copy and complete this stage of the radioactivity decay series shown below, giving values for a, b, c and d, and naming the element X. **5**

$$^{234}_{92}U \longrightarrow\ ^{a}_{b}X +\ ^{c}_{d}\alpha$$

(*b*) The following graph shows how the effective dose equivalent rate due to background radiation varies with height above sea level.

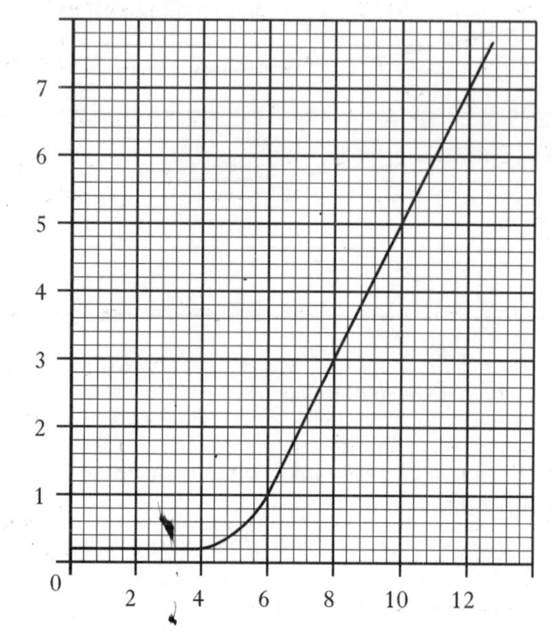

(i) Name **two** sources of background radiation.

(ii) The graph shows that there is an increase in effective dose equivalent rate at altitudes greater than 4 km. Suggest a reason for this increase.

(iii) An aircraft makes a 7 hour flight at a cruising altitude of 10 km.

(A) Calculate the effective dose equivalent received by a passenger during this flight.

(B) A regular traveller makes 40 similar flights in one year and spends the rest of the year at sea level.

Calculate the effective dose equivalent of background radiation received by this traveller in that year. **8**

(13)

[END OF QUESTION PAPER]

Printed by Bell & Bain Ltd., Glasgow, Scotland.